Towards a European Forest Information System

Authors of This Volume

Andreas Schuck, European Forest Institute
Tim Green, European Forest Institute
Gennady Andrienko, Fraunhofer-Gesellschaft zur Förderung der angewandten Forschung e.V. (FhG/AIS)
Natalia Andrienko, Fraunhofer-Gesellschaft zur Förderung der angewandten Forschung e.V. (FhG/AIS)
Alex Fedorec, University of Greenwich
Aljoscha Requardt, University of Hamburg (UHH), Institute for World Forestry
Tim Richards, Conservation Technology Ltd
Roger Mills, Plant Sciences Library, Oxford University
Eero Mikkola, IUFRO
Risto Päivinen, European Forest Institute
Michael Köhl, University of Hamburg (UHH), Institute for World Forestry
Jesus San-Miguel-Ayanz, Joint Research Centre, Institute for Environment and Sustainability

VOLUME 20

Towards a European Forest Information System

European Forest Institute Research Report 20

By

Andreas Schuck
Tim Green
Gennady Andrienko
Natalia Andrienko
Alex Fedorec
Aljoscha Requardt
Tim Richards
Roger Mills
Eero Mikkola
Risto Päivinen
Michael Köhl
Jesus San-Miguel-Ayanz

EUROPEAN FOREST INSTITUTE
JRC
EUROPEAN COMMISSION

TUTA SUB AEGIDE PALLAS · 1683 ·

BRILL

LEIDEN • BOSTON
2007

This report was prepared for the project "Network for a European Forest Information Service (NEFIS)", which was carried out as an accompanying measure in the Quality of Life and Management of Living Resources Programme of the European Commission (contract number QLK5-CT-2002-30638). The content of this publication is the sole responsibility of the authors and does not necessarily reflect the views of the European Union or the European Forest Institute.

This book is printed on acid-free paper.

Library of Congress Cataloging-in-Publication Data

A C.I.P. record for this book is available from the Library of Congress.

ISSN: 1238-8785

ISBN-13: 978 90 04 16150 4
ISBN-10: 90 04 16150 3

PRINTED IN THE NETHERLANDS

Table of Contents

Foreword

Forest sector experts and professionals often complain about the prejudices and misunderstandings about forest issues. One cause of this is surely the difficulty of finding reliable, well structured information. The idea of a European forest information system to reduce this problem has been pursued for some time: the present report represents a further step forward.

Conventions and political processes both at national and international levels have formally recognized the need to access and use environmental and thus forest information. The regulation (EC) No. 1615/89 as of 1989, which expired in 2002, stated that the European Commission (EC) should set up a European Forest Information and Communication System in order to address the need for sound forestry information at the European level.

From 1989 until the present, numerous research activities and development projects have been or are being undertaken to further progress towards an operational European forest information system. The "Network for a European Forest Information Service (NEFIS)" project, ongoing between 2003 and 2005, had as its main goal to contribute to this development.

The activities of the NEFIS project included the exploration of an overall forest information system architecture based on existing structures at national, EU and international levels. It gave explicit attention to the development of harmonized standards and procedures for providing metadata. The project also demonstrated options of remote access to data hosted at different locations. The project included various actors of a European forest information system – namely data providers and users, the IT community, and terminology experts – into its consortium. Such an approach made it possible to familiarize all actors with the concept of a European forest information system, in particular providing them with a better understanding of the opportunities and limitations.

The NEFIS results have found their use in a subsequent activity supported by the EC Joint Research Centre where an operational version of a European forest information system is currently being developed. The 'European Forest Information and Communication Platform' will be welcomed by the NEFIS partners as well as many actors dealing with and in need of forestry data and information either as data providers or users.

I would like to take this opportunity to express my special thanks to the dedicated NEFIS project partners, the work package leaders, invited experts and reviewers without whom this project would not have achieved its challenging goals. Despite many political and financial obstacles, the authorities of the EU have consistently worked to make more and better information available to governments, the research community and stakeholders. A very special thanks goes to my colleagues of the Steering Committee Tor-Björn Larsson, Stefanie Linser and Jesús San-Miguel-Ayanz who have made my task as a chairman both easy, and pleasant by sharing their expertise and valuable advice both with the NEFIS partners and myself.

Geneva, 23 July 2007
C.F.L. Prins
Chief, UNECE/FAO Timber Section
NEFIS Steering Committee Chairman

Acknowledgements

The coordinator of the Network for a European Forest Information Service (NEFIS – contract number QLK5-CT-2002-30638) project would like to thank all members of the consortium, and numerous others for their contribution. The project was carried out as an accompanying measure in the Quality of Life and Management of Living Resources Programme of the European Commission.

Chapter 1 of this report provides a history and background to the NEFIS project and was written by Andreas Schuck with contributions from Risto Päivinen, Jesús San-Miguel-Ayanz, Eero Mikkola and Tim Green.

Chapter 2 describes a basic architecture for a European forest information system and is based on the report concerning the NEFIS UML model (deliverable 9) written by Alex Fedorec and Tim Richards with support from Keith Rennolls and Moh Ibrahim.

Chapter 3 discusses the issues concerning metadata and draws on the metadata standards and keyword lists report (deliverable 3) of the NEFIS project, and in particular the interoperability section of that report written by Roger Mills, the controlled vocabulary section by Gillian Petrokofsky, Renate Prüller and Eric Landis, and the evaluation report (deliverable 10) by Aljoscha Requardt and Michael Köhl.

Chapter 4 describes the advanced demonstrator developed within the NEFIS project and was written by Tim Richards, Gennady Andrienko, Natalia Andrienko, Andreas Schuck and Tim Green. Aljoscha Requardt and Michael Köhl provided material concerning the evaluation of the demonstrator.

Chapter 5 draws on the reports and contributions from all of the NEFIS consortium and the working groups of the final symposium. Specific contributors to this chapter were Andreas Schuck, Tim Green, Tim Richards, Gennady Andrienko, Natalia Andrienko, Aljoscha Requardt, Michael Köhl and Risto Päivinen.

NEFIS consortium.

- European Forest Institute (EFI): Risto Päivinen, Andreas Schuck, Tim Green, Simo Varis, Sergey Zudin, Jo Van Brusselen. Tim Richards (Conservation Technology Ltd) was also a member of the EFI team.
- Joint Research Centre, Institute for Environment and Sustainability (JRC IES): Jesús San-Miguel-Ayanz; Sten Folving.
- International Union of Forest Research Organizations (IUFRO): Eero Mikkola, Binh Thanh Nguyen, Peter Mayer, Nils Bruun de Neergaard, Renate Prüller. Eric Landis (Natural Resources Information Management), Gillian Petrokofsky (CAB International), Roger Mills (Oxford University) contributed significantly with their expertise to the work of the IUFRO team.
- University of Hamburg (UHH), Institute of World Forestry: Michael Köhl, Aljoscha Requardt.
- Fraunhofer-Gesellschaft zur Förderung der angewandten Forschung e.V. (FhG/AIS): Natalia Andrienko, Gennady Andrienko, Hans Voss.

- Finnish Forest Research Institute (Metla: Jarmo Saarikko, Mika Mustonen, Helena Mäkelä, Martti Aarne, Erkki Tomppo.
- Inventaire Forestier National (IFN): Marie-Claire Gúero, Jean Wolsack.
- Danish Centre for Forest, Landscape and Planning (KVL): Vivian Kvist Johannsen, Jens-Peter Skovsgaard.
- Sveriges Lantbruksuniversitet (SLU): Göran Kempe, Per Nilsson, Göran Ståhl.
- University of Greenwich (UG): Keith Rennolls, Moh Ibrahim, Alex Fedorec.
- Regione Dell'Umbria (RU): Francesco Grohmann.
- Italian Academy of Forest Sciences (AISF): Gherardo Chirici; Marco Machetti.
- Forest Technology Centre of Catalonia (CTFC): Marc Coromines, Gloria Domínguez-Torres, Monica Bori, Iolanda Domenjó.

We would also like to thank the Federal Research Centre for Forestry and Forest Products (Richard Fischer, Martin Lorenz, Volker Mues), and the Hungarian State Forest Service (Zoltan Kovacs, Pal Kovacevics, Petter Kottek) for supplying datasets and commenting on various aspects of the project.

Thanks also go to the members of the Steering Committee with its chairman Kit Prins (UNECE/FAO), Tor-Björn Larsson (EEA), Stefanie Linser (EEA National Focal Point, Umweltbundesamt, Austria) and Jesús San-Miguel-Ayanz (representing EC).

Finally we thank the reviewers of this report who provided us with valuable constructive reviews that have enabled us to improve the manuscript considerably – Robin Quenet (Project Manager of the Canadian Council of Forest Ministers National Forest Information System) and Adrian Lanz (Head of Land-Resource Assessment research group at the Swiss Federal Institute for Forest, Snow and Landscape Research).

Acronyms and Abbreviations

CBD	Convention on Biological Diversity
CCD	Convention to Combat Desertification
CPF	Collaborative Partnership on Forests
CSD	Commission on Sustainable Development
DC	Dublin Core
DCMES	Dublin Core Metadata Element Set
DCMI	Dublin Core Metadata Initiative
EC	European Commission
EFICP	European Forest Information and Communication Platform
EFICS	European Forestry Information and Communication System
EFIS	European Forest Information System
ENFIN	European National Forest Inventory Network
EU	European Union
FAO	Food and Agricultural Organization of the United Nations
FCCC	Framework Convention on Climate Change
FRA	Forest Resource Assessment
GFIS	Global Forest Information Service
GIS	Geographic Information System
IIASA	International Institute for Applied Systems Analysis
INSPIRE	INfrastructure for SPatial InfoRmation in Europe
ISO	International Organization for Standardization
IUFRO	International Union of Forest Research Organizations
JRC	Joint Research Centre of the European Commission
MCPFE	Ministerial Conference on the Protection of Forests in Europe
NEFIS	Network for a European Forest Information Service
NGO	Non-Governmental Organization
PSI	Public Sector Information
RD	Resource Discovery
RPC	Remote Procedure Call
RSD	Remote Search Demonstrator
SDI	Spatial Data Infrastructure
SOA	Service Oriented Architecture
SOAP	Simple Object Access Protocol
TBFRA	Temperate and Boreal Forest Resource Assessment
UML	Unified Modelling Language
UNECE	United Nations Economic Commission for Europe
UNEP	United Nations Environment Programme
UNFCCC	United Nations Framework Convention on Climate Change
VTK	Visualization Toolkit
WCMC	World Conservation Monitoring Centre
WWF	WWF: Global environmental conservation organization
XML	eXtensible Markup Language

Executive Summary[1]

Information needs

European forests are playing an increasingly important role in terms of provision of ecosystem services – such as production of timber and non-wood forest products; as a bioenergy source; provision of recreation; protection of water and soil; a sink for atmospheric CO_2; provision of habitat and maintenance of biodiversity. Aims, interests and needs with respect to forests will thus vary according to the different stakeholder groups, and consequently the needs for specific information will also vary.

The need to improve both use of and access to environmental information was formally recognized as a priority in 1992 by the United Nations Conference on Environment and Development, UNCED (Agenda 21, Chapter 40). A European Union (EU) regulation was put in place which encouraged the coordinated collection and standardization of European forest sector information prior to the UNCED. The regulation (EC) No. 1615/89 as of 1989 (expired in 2002) stated that the European Commission (EC) should set up a European Forest Information and Communication System (EFICS) in order to address the need for sound forestry information at the European level[2]. The main goal of this regulation was that data and information concerning the forestry sector and its development should be collected, standardized and processed. Existing data should be utilized in compiling particular statistics by the European Communities statistical office and information from Member States and other available and accessible databases both at the national and international levels. Based on this regulation a first prototype of a potential European forest information system was developed within a project for the EC Joint Research Centre during 2000–2002 (European Forest Information System – EFIS project: 17186-2000-12 F1ED ISP FI). A follow-up project was implemented within the 5th Framework Programme project entitled "Network for a European Forest Information Service", NEFIS (QLK5-CT-2002-30638). The NEFIS project ran from 2003 to 2005, and its outcomes are the subject of this report. The NEFIS results are inputs to a subsequent activity supported by the EC Joint Research Centre to develop a European Forest Information and Communication Platform (EFICP).

The NEFIS project

The project explored an overall information system architecture based on existing data reporting structures at national, EU and international levels. The project gave explicit attention to the development of harmonized standards and procedures for providing metadata. The NEFIS consortium included data providers and users, the IT community, and terminology experts. Emphasis was placed on ensuring that data providers and users were made familiar with the concept of a European

1 Note on the use of terms in this summary and in the text of this book. The phrase "European forest information system" is used to refer to a generic information system. "EFIS" is used to refer specifically to the European Forest Information System project and the outcomes of the EFIS project.

2 EC, 1989. Regulation (EC) No 1615/89 of 29 May 1989. Establishing a European Forestry Information and Communication System (EFICS). Official Journal, L 165 , 15/06/1989 P. 0012 – 0013.

forest information system whilst developing a framework for supporting potential data providers at various spatial scales. The data providers provided datasets and the associated metadata according to a specified metadata schema. Building on the work carried out for the EFIS project, an advanced demonstration version of a European forest information system was developed. The datasets were made available through the demonstrator. The various components necessary for establishing the demonstrator as well as its functionalities were subject to an extensive evaluation process.

Elaboration of an overall system architecture

A first step in the design of the overall system architechture was the need to understand the context in which it would be used. A number of characteristic European forest use cases were defined – i.e. brief descriptions of data/information collection processes and their various actors and requirements. Having mapped different data collection and dissemination activities the overall architecture of a forest information system could be designed. In essence the system should be based on a distributed, 'publish-subscribe' pattern of a service-oriented architecture. Such a structure is simple and based on collaborations between a 'publisher' of a service and a 'subscriber' who locates and uses those services via a directory. The publisher maintains ownership and control of published information at their own premises. Visualization, analysis and other processing of information are done at the subscriber node. Once the information has been processed, the roles may be switched, and the subscriber may become publisher. The server of such an information system will need to exploit Web services interoperability technologies to: provide the glue service that enables data discovery; provide gateway services that enable users to access data; and act as a central tool and resource repository to enable data analysis (see Figure 16, section 2.4.5).

Metadata

The extraction and analysis of data from a variety of sources involves a number of stages which include: (a) the identification of appropriate sources; (b) their evaluation for relevance and reliability; (c) the actual process of extracting data of interest to the user; (d) manipulation using tools appropriate for the required purpose; (e) interpretation of the compiled results; and (f) their presentation to the appropriate audience(s). This requires that background information should be explicitly stored with the data and be retrievable to a user at any point of the knowledge generation process. That is the purpose of metadata. Therefore, one of the tasks in the NEFIS project was to employ metadata to provide a clear description of information resources, and thus support interoperability in a standards-led process. The datasets submitted for the NEFIS project, however, were too heterogeneous to enable them to be made interoperable. Thus the development of the NEFIS metadata focused on descriptive metadata which is sufficient for the purposes of cataloguing and identifying relevant datasets. The outputs are human-readable information on the content of those datasets. The NEFIS metadata itself is not sufficient for the purpose of automated cross-searching of selected datasets and compilation of combined data for submission to processing tools without considerable user intervention.

For the purposes of NEFIS a recognized metadata standard – the Dublin Core Metadata Initiative) with its Dublin Core Metadata Element Set (DCMES) was adopted as the starting point. DCMES consists of 15 main metadata elements, and these elements can be qualified (i.e. refined and added to). The metadata schema was developed to include more details to describe spatial information and to address in more depth the issue of data quality. The addition of an element entitled 'reference system' provides a description of spatial information and closely follows the ISO 19115 standard for metadata for geographic information. Data quality which was seen by the data providers as a crucial aspect was tackled by adding a section entitled 'quality report' as an element refinement under the DCMES element 'description'. The quality report included information on definitions, availability of data collection and data processing guidelines, sampling methods and collection mandate.

Controlled vocabulary for metadata

NEFIS initiated the process of building a controlled vocabulary for the NEFIS data and information resources. The vocabulary was used by the data owners to catalogue the NEFIS data and information sources, and gives information seekers the means to identify and retrieve NEFIS data by using a common list of consistently applied terms. From the beginning of the NEFIS project it was clear that development of a full controlled vocabulary for European forestry was well above the resources available within the project.

Through cooperation of the NEFIS partners an attempt was made to explore different approaches and possible solutions to develop a controlled vocabulary. The work followed closely from existing recognized vocabularies such as the CAB Thesaurus, the FAO AgroVoc, the (US) National Agricultural Library Thesaurus, and also vocabularies used by the project partners where these were available. NEFIS partners identified a set of 'themes' reflecting the datasets which they would provide for the project. Examples are forest inventory, forest health, silviculture, forest products and trade flows, maps and geo-referenced data and field experiments. The 'NEFIS themes' were not seen as a new classification within the forestry domain, but as a contribution to ongoing activities in vocabulary and thesauri development. In particular the processes of establishing the NEFIS themes and the corresponding controlled vocabularies or subject keyword lists under these themes were investigated using a cooperative consultation process between data providers, users, and ontology and library experts. The project partners were then asked to submit keyword terms for the themes for which they could provide expertise. The resulting keyword lists were variable in the level of detail and scope. In parallel with this effort, terminology experts identified existing relevant keyword lists that could be used as a base for the NEFIS vocabulary. Both activities were combined resulting in a vocabulary for the themes of the NEFIS project. The vocabulary was used by data providers to describe their NEFIS datasets under the DCMES metadata element 'subject'.

Test data used in the NEFIS project

The NEFIS partners were asked to provide representative datasets from their organizations which would be used in testing the components of the forest information system demonstrator (i.e. the resource discovery and data exploration tools). The datasets were described using the DCMES metadata elements and prepared for the use of distributed data retrieval, analysis and visualization. The datasets consisted of national forest inventory data, international data on forest resources, forest products and trade, socio-economics and forest condition. They also included data on rural development and land use, forest fires and forestry experiments. The complete list of data used in the NEFIS project is presented in Table 12 (section 4.3.2).

Data analysis
Resources – i.e. NEFIS data tables/databases – were made available for downloading from one or more sources (partner organizations) and locations (distributed data retrieval). The data were linked to data exploration tools which are one of the forest information system demonstrator components. The tools allow further processing and visualization of the linked data. This could be done in real time from any given location with access to a Web browser. A so-called visualization toolkit (VTK), was first designed for the EFIS project and further developed within NEFIS. The VTK is based on the CommonGIS system for thematic mapping and exploratory data analysis. The capabilities of the VTK for the NEFIS project were intended to reach beyond the mere generation of maps and graphs. Particularly, the VTK is useful for the comprehensive exploration of new, previously unknown data. Such exploratory data analysis builds on the unbiased examination of data in order to detect and describe patterns, trends, and relationships in the data. Such options will constitute a main component of an operational system in which different actors can perform a variety of operations ranging from simply displaying data to more comprehensive data checks and data exploration. The VTK is one example of a data analysis tool. For the NEFIS demonstration, the datasets had to be prepared specifically for use with the VTK. If standards for the harmonized metadata descriptions of dataset structures were developed and detailed metadata descriptions of dataset structures became available, then tools such as the VTK could be developed in such a way that the need for human intervention is lessened, or even removed.

The forest information system demonstrator
One of the important developments in the NEFIS project was the embedding of the actual operation of the information system demonstrator to run within a network of distributed datasets. This implied that the datasets were hosted by the individual data providers and can be retrieved by a user from those locations. For the NEFIS demonstrator two approaches for data retrieval were distinguished. Both approaches were to allow further exploration and visualization of the data once they were retrieved.

1. Retrieval of an individual dataset from one remote server location: data physically located on the data server of a NEFIS partner
2. Retrieval of data (forest inventory data) from multiple countries located on remote servers with one query request. This means that the data are retrieved from physically different locations (servers) simultaneously and compiled into one table for further display and analysis.

The approach described under bullet number 2 is in particular interesting. It provides a mechanism to extract data from a standardized set of distributed data sources, collate the data, present it in a table and convert the data into a form that can be used by the VTK. At least two actors are involved in this process: (a) data users who want to discover, extract and use data; and (b) data providers who have data to share. The mechanism of this approach is based on a 'Remote Procedure Call' (RPC) which allows software running on disparate operating systems, running in different environments to make procedure calls over the Internet. The demonstration RPC in NEFIS does still imply that a user identifies a particular metadata record in the metadata catalogue from which the RPC can then be initiated. Ideally an operational European forest information system would allow a user to specify a search string, which prompts the direct extraction of target data simultaneously from one or more sources (servers). The NEFIS example RPC case uses forest inventory data for forest area located on different servers. The approach and the results of a query are illustrated in Figure 26 and Figure 36, respectively.

Recommendations and challenges
The activities of the NEFIS project and its predecessor projects, addressed in particular a framework for a European forest information system including the investigation of potential standards, a generic system architecture, and various technical options. During the NEFIS project the system components such as resource discovery and data exploration tools of an EFIS were further developed. There are, of course, numerous challenges that remain. The outcomes of the NEFIS project, including a set of recommendations, may thus assist in advancing towards building an operational forest information system.

1. Ideally, an operational forest information system should avoid development of proprietary technologies and use standardized components that allow the capture and storage of dataset information necessary to make interoperability feasible. Information systems and technologies will evolve, but, if established standards are adopted, data migration should be relatively straightforward. This implies that clear decisions must be made concerning the standards needed to support information retrieval and interoperability. Those decisions must be made in the light of consensus in other subject areas, and possible EU requirements. NEFIS has addressed the issue of standards to the degree possible within the project framework. A non-exclusive list of required metadata components were suggested accompanied by examples of the standards or technologies that could be adopted when building an operational information system. They include: top level metadata formats, data syntax, registration of semantics of shared data elements and semantic structures, interaction at systems interfaces, compatibility/adoption of standards and metadata interoperability. In any follow up activities it is recommended that discussions on standards should involve all interested parties, including users at a broad scale. This could best be achieved by a specialized working group, preferably including specialists who are familiar with interoperability issues, and representatives from other communities within the EU who are aware of emerging regulatory frameworks. This 'community of interest' (COI) could then determine future working structures towards a European forest information system.

2. Current activities in the field of terminology work point towards the development of ontologies for the semantic web. Sound development of ontologies needs to be based on terminology, thesauri and classifications including definitions, relationships and structure. Transforming existing thesauri into ontologies can yield increased precision of semantics particularly for automated information retrieval purposes. Based on the experiences gained during the activities of vocabulary development in NEFIS, some observations were made on how to advance vocabulary development in the forestry domain. Ongoing vocabulary activities should be coordinated within a 'Multilingual Forestry Ontology Project', with strategic links to ongoing ontology framework projects such as, for example, the FAO Agricultural Ontology Service (www.fao.org/agris/aos/). The establishment of an editorial advisory group, comprising subject experts and information specialists, could take a role in collating and organizing terminology to maintain the forestry ontology. NEFIS would encourage the use of the Global Forest Decimal Classification (GFDC) as part of the collection of metadata related to forestry resources. The GFDC replaces the former Forest Decimal Classification (FDC) and Oxford System for Decimal Classification for Forestry (ODC) and was published in 2006.

3. For an architectural design of a European forest information system it is recommended by NEFIS that any proposed system be an evolving collaboration of communicating systems developed iteratively. It should employ a component oriented implementation exploiting the current extensive portable web and open-systems technologies. The forest information system should be based on a partitioned scalable architecture within an extensible service-oriented framework. To ensure the exploitation of highly cohesive, lowly coupled components within what is essentially a distributed heterogeneous system requires a common interchange language and middleware with agreed terminology. This role can be filled by metadata representing the 'glue' that ensures the semantic interoperability (i.e. agent dialogue) of the system. NEFIS further recommends as part of an overall system structure, the development of effective exploration and visualization tools which should be available within a tool repository component. Such tools should be developed so that tailored needs and requirements of different user groups can be served.

4. Existing international reporting processes will benefit by linking national datasets in one European service, enabling simpler, faster, more flexible and more reliable access to data (in the cases where data are comparable). International reporting requires that national data are comparable to allow holistic assessments about status and trends of relevant concerns. Delivering national data according to international standards often requires data transformation. The development of 'conversion tools' in the form of software or statistical scripts which can be implemented into national databases to convert 'national data' into 'international data' are seen as a feasible way forward. However, such tools are to date not very far developed. Once available, such conversion tools could become an integral part of a European forest information system. The system could then more effectively contribute to minimizing the individual national reporting burden.

1. Introduction

1.1 The need for (forestry) information

A first question one may ask is why we need information. Further to this, questions may arise on who needs information, how and where to access it, and in which form. In this report we will introduce the reader to these issues based on the outcomes of a number of previous research activities. They are necessary background to better understand the work that has been performed in the 5th Framework Programme Accompanying Measure project Network for a European Forest Information Service, NEFIS (2003–2005) which is the subject of this report. The NEFIS project itself focussed mainly on issues of access to and provision of forestry information. Although the questions of "who needs information?" and "in what form do they want it?" are introduced in this report, they were not within the scope of the NEFIS project. The report also incorporates a vision towards the establishment of a so-called European Forest Information and Communication Platform which may be seen as one component of a data portal for the European Union.

Member States of the EU are obliged to facilitate re-use of public sector information. Member States shall ensure that: (1) "where the re-use of documents held by public sector bodies is allowed, these documents shall be re-usable for commercial and non-commercial purposes"; and (2) "practical arrangements are in place that facilitate the search for documents available for re-use, such as asset lists, accessible preferably online, of main documents, and portal sites that are linked to decentralized assets lists" (EC, 2003).

Therefore, there is a need (and an obligation in particular for the EU and its Member States) to provide systems and services that enable the re-use of this data/ information.

The INSPIRE (Infrastructure for Spatial Information in Europe) Directive 2007/2/EC (EC, 2007) acknowledges that detailed spatial information is available in Europe, but that widespread access to and use of this information is problematic – the main problems relating to data gaps, missing documentation, incompatible spatial datasets and services due to varying standards and barriers to the sharing and re-use of spatial data (EC, 2004c). INSPIRE is taking a leading role in establishing a Spatial Data Infrastructure (SDI) for Europe. INSPIRE will follow ISO standards, and ISO in turn is closely cooperating with the Open Geospatial Consortium. The current state of development of SDIs within EU Member States is varied, but strides forward are clearly being made by Member States – 24 out of 25 Member States agreeing or agreeing in part with the statement that "metadata are produced for a significant fraction of geodatasets of reference data and core thematic data"; 17 Member States agreed or agreed in part with the statement that "one or more standardized metadata catalogues are available covering more than one data producing agency"; and 16 agreed or agreed in part "there is a coordinating authority for metadata implementation at the level of the SDI" (Vandenbroucke, 2005).

Data, information and knowledge are terms that are frequently used for overlapping concepts. In general terms 'data' are collections of facts represented in a language (such as numbers, characters, images, or other methods of recording on a durable medium) that is readable by humans or machines. Data on their own carry no meaning (ICHNET, 2005). Spatial data are data that refer to a specific location.

Information however can be defined in a number of different ways. It can be a message, in the form of a document or an audible or visible communication, meant to change the way a receiver perceives something and to influence judgement or behaviour (ICHNET, 2005). It can also be defined as data that makes a difference (Davenport and Prusak, 2000) or it can represent patterns in data (O'Dell and Jackson Grayson Jr., 1998). It is the linkage of data (syntax) and the associated meaning (semantics) (Köhl, 2006).

Knowledge can be defined as "what is known by perceptual experience and reasoning" (ICHNET, 2005). Knowledge can either be gained through experiential knowledge (O'Dell and Jackson Grayson Jr., 1998), systematic investigation but also through deduced cognition (Köhl, 2006).

If we now take an example: There are about 200,000 km^2 of forest in a EU country. We can think of 200,000 km^2 as data. A statement that "this EU country's forest land area has increased by more than 16,000 km^2 since the 1960s" can be regarded as information; "The increase in forest land area since the 1960s is a result of afforestation of agricultural and other lands as well as intensive forest improvement efforts" is knowledge. This may have been derived through systematic observations or investigations. When now combining information/knowledge from different domains (e.g. economical and technological aspects, environment, etc.) forest managers may decide to further increase, decrease or keep the forest area stable. In this way combined information/knowledge may lead to a specific action, development of methods or technology to reach a certain aim thus becoming know-how which may be of interest to other parties having similar developments (Schuck et al., 2005). In this context, one should keep in mind that the information and knowledge are only as reliable as the underlying data from which they were derived.

1.2 Who needs (forest) information?

Forests play an important role in Europe in terms of provision of ecosystem services – such as production of timber, non-wood forest products; provision of recreation and aesthetic facilities; protection of water and soil; a sink for atmospheric CO_2; provision of habitat and maintenance of biodiversity. The multiple functions of the forest are receiving a large amount of attention in forest management and landscape planning both at the policy and the decision making levels. Various groups and organizations at international, EU, national, regional and local scales have different aims, interests and needs with respect to forests, and consequently have very specific information needs.

Therefore, the success of a system for providing forestry information will depend strongly on the degree to which the information needs of potential users will be satisfied. Thus the demand, as well as the supply, side of information has to be taken into account. Comparing both demand and supply allows identification of current discrepancies in information needs and information availability, and also of the potential components of a system for satisfying the identified user needs.

Information to be provided by an information system on forests may vary according to the purpose and the scale for which it is used. Information needs that are very task specific will require a set of attributes with large thematic and spatial degree of detail (e.g. information used for forest management purposes). Other information of a more strategic and integrative nature will require more aggregated information both in thematic and spatial terms. The need for aggregated information is more evident at the European level as compared to specific issues of interest at national, regional or even local scale (Tables 1 and 2).

Table 1. Potential user groups for an information system on forests (no specific order; adapted from: (JRC, 2002b; Päivinen et al., 1998; Päivinen and Köhl, 2005).

National perspective	European/international perspective
• Ministries of Agriculture and Forestry, Ministries of Environment (and other ministries as appropriate) • State forestry organizations; • Forest industry associations or industry sector associations • Associations of private forest owners • Nature Conservation/Environmental organizations • Main forest research organizations in the country • Members of the EU Standing Forestry Committee	• International processes (with emphasis on European aspects e.g. MCPFE, UNFCCC; CBD) • European Commission • European Parliament • Eurostat • European Environment Agency • Joint Research Centre • FAO, UNECE/FAO • IIASA • International expert and advisory groups/panels • European Associations of Forest Owners • Forest Industry associations • WWF; UNEP-WCMC • European Centre for Nature Conservation

Table 2. List of forestry information needs identified during the EFICS and EFIS studies (adapted from (JRC, 2002b; Päivinen et al., 1998; Päivinen and Köhl, 2005).

Information needs	
• Forest resources • Production, trade and utilization • Socio-economic information • Environmental and biodiversity aspects • Land use • Supply of non-wood goods and services • Climate changes	• Forest policy and legislation • Forest condition • R&D activities • Sustainable development and management • General interest on forest related issues

1.3 How and where to access forest information?

1.3.1 Introduction

Improving access to environmental information, including forests, was formally recognized as a priority by the United Nations Conference on Environment and Development as early as in 1992 when it stated in Agenda 21, Chapter 40:

> *Existing national and international mechanisms of information processing and exchange, and of related technical assistance, should be strengthened to ensure effective and equitable availability of information generated at the local, provincial, national and international levels... (UNCED, 1992).*

Further the central role of 'electronic information systems' was also noted very clearly in Chapter 40:

> *Countries, international organisations, including organs and organisations of the United Nations' system, and non-governmental organisations should exploit various initiatives for electronic links to support information sharing, to provide access to databases and other information sources, to facilitate communication for meeting broader objectives, such as the implementation of Agenda 21, to facilitate intergovernmental negotiations, to monitor conventions and efforts for sustainable development, to transmit environmental alerts, and to transfer technical data (UNCED, 1992).*

The Intergovernmental Panel on Forests in 1997 restated the importance of gaining access to information from a forestry perspective:

> *The Panel emphasised the need to review and improve information systems. Attention should be given to world-wide access to information systems that would encourage effective implementation of national forest programmes, increased private-sector investment, efficient development and transfer of appropriate technologies, and improved co-operation (CSD, 1997).*

Presently there are a large variety of information sources existing on forests at the international, European, national, regional and local level. Examples at the international and pan-European levels are the Forest Resources Assessments carried out by FAO and UNEC/FAO which started as early as the 1950s (FAO, 2006; Gold, 2003; UN/FAO, 2001; UNECE/FAO, 2000), the FAOSTAT which includes data on forest products and trade (FAOSTAT, 2006), the UNECE/FAO's timber market and price statistics (www.unece.org/trade/timber/), the Liaison Unit services of the Ministerial Conference on the Protection of Forests in Europe, MCPFE (MCPFE, 2003b), the Eurostat Forestry Statistics (Eurostat, 2000; Eurostat, 2003), the European Environment Agency's European Environment Information and Observation Network, EIONET (www.eionet.eu.int/) and at national level various forest inventory reports and forestry statistics (EC, 1997; Schelhaas, 2003; Schelhaas et al., 2006).

However, information on forests is still quite scattered and partly incomplete. The information does not necessarily cover all potential application fields where data are needed or they are not available at adequate depth – for example, to fully comply with international reporting commitments. There are also evident differences between countries in the intensity and timeframe of collecting data, their methodological approaches and data processing procedures (EC, 1997; FAOSTAT, 2006; Päivinen and Köhl, 2005; Wardle et al., 2003). This means that in certain cases even basic comparison of statistical information can become rather difficult or even impossible. There are, however, numerous initiatives at the international level dealing with the issue of definitions and harmonization (COSTE43, 2004; FAO, 2004; MCPFE, 2003a) and comparable data for key variables have been or are being made available.

Despite the ongoing initiatives regarding data collection and harmonization there are key issues that still need to be tackled. These include: (1) the reduction of the heterogeneity of the various data sources; and (2) the development of a reliable forest information system to compile, process, analyze and disseminate available information. Already in the year 1989 the regulation (EC) No. 1615/89 stated that the European Commission should set up such a European Forest Information and Communication System (EFICS) in order to address the need for sound forestry information at the European level (EC, 1989). The main objective of the EFICS was to collect, co-ordinate, standardize and process data concerning the forestry sector and its development. Existing data should be utilized in compiling particular statistics by the European Communities statistical office and information from Member States and other available and accessible databases both at the national and international level. This particular regulation has expired in 2002. The European Commission implemented a comparative study entitled 'Study of European Forestry Information and Communication System – Reports on forestry inventory and survey systems' in 1997 as it recognized the lack of information on the various systems for collecting, processing and presenting forestry data (EC, 1997). The study revealed the diversity of inventory and sampling procedures, nomenclature, modelling, analysis techniques, inventory cycles and organizational structures in the participating countries.

With the help of the EFICS study it was possible to illustrate both alternative options to improve the comparison and reliability of forest statistics and give a first vision on how a reporting system could be established (Figure 1). The reporting system vision paid due respect to principles of utilizing existing, national forestry data, and other official national and international data collection initiatives as well as respecting the the rights of data owners and visible recognition of the original sources. It also took into consideration the need to provide an adequate amount of harmonized data for EU data collection, response to data needs relating from international commitments and trans-boundary research. Therefore the most promising option within the EFICS study was seen as A1→B1→C3 (Figure 1).

The activities to establish a Global Forest Information Service (GFIS) was initiated in 1998 at the International Consultation on Research and Information Systems in Forestry (ICRIS) held in Gmunden, Austria (Päivinen et al., 1998). Today GFIS is an internet-based metadata service which aims to enhance access to all types of forest information, ensuring that it is accessible to governments and to

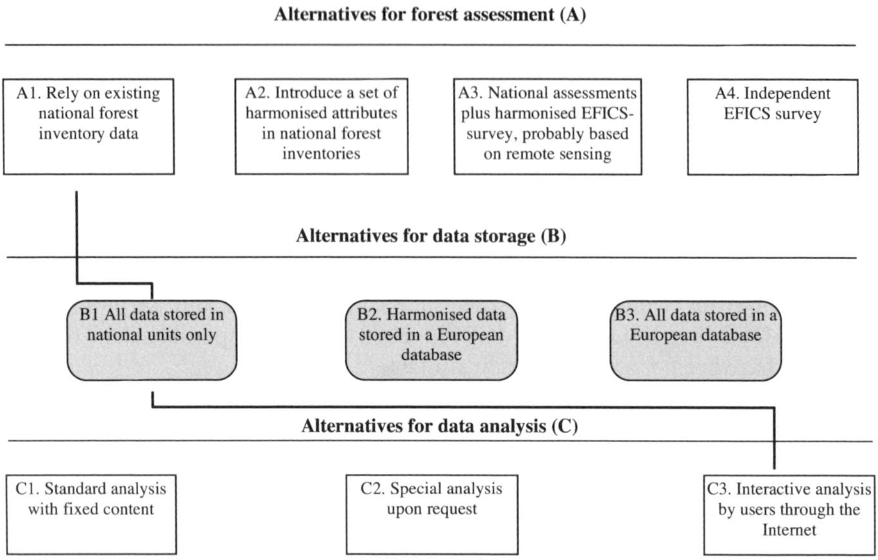

Figure 1. Options for the development of the European Forestry Information and Communication System as resulting from EFICS (EC, 1997).

all stakeholders, including researchers, forest managers, NGOs, community groups and the public at large. It aims to contribute to an improved understanding of complex forest-related issues, to better decision-making and more informed public engagement in forest policy and forest management at all levels (GFIS, 2005). GFIS is an initiative of the Collaborative Partnership on Forests (CPF), an innovative interagency partnership of 14 major forest-related international organizations, institutions and convention secretariats. The actual development of GFIS is implemented under the umbrella of the International Union of Forest Research Organizations (IUFRO). The activities of GFIS are of high relevance in the context of developing an operational European forest information system.

1.3.2 The European Forest Information System (EFIS) Demonstrator

Based on the regulation (EC) No. 1615/89 and the outcomes of the EFICS study of 1997 a next step in the process of building an information system was to look into the technological possibilities of addressing the main aims of the regulation. The project "European Forest Information System (contract no. 17186-2000-12 F1ED ISP FI) had as its aims to establish an objective, comprehensive, innovative, user-driven information system built on an architectural design and technological components that would guarantee both continuity of operation and the options of improvement and further development (JRC, 2002a; Kennedy et al., 2004; Schuck et al., 2005; Schuck et al., 2004). The EFIS was to fulfil the basic requirements of:

a) *Objectivity and compatibility*
EFIS was to be based on officially recognized forest information and assuring high data quality and transparency of the information. The system should be constructed so as to allow for various data types and data formats.
b) *Innovativeness*
The system was to be reliable, flexible and user friendly based on technology that is available to users conveniently and at low cost. It should fulfil the requirement of platform independence and embrace recognised and well established standards.
c) *A system for the user*
It was a requirement that the system should be tailor-made to serve various user groups according to their skills and technical facilities.
d) *Continuity*
The system would need to demonstrate that it could be regarded as open-ended, interactive and cost effective in order to represent a permanent and consistent source of forestry information for the European Union member states.

The work undertaken in EFIS was three-fold: (1) it included the evaluation of technical alternatives for an information system demonstrator; (2) it explored innovative ways of querying information, extracting, processing and visualizing data; and (3) it touched on the issues of data quality and data access protocols.

The EFIS demonstrator architecture is shown in Figure 2 (simplified). The steps for an information searcher are at first to search for desired data/information through a metadata service. This can either be done through a free text search or using keywords via a browse function (resource discovery). Once the required metadata is identified a web address (or identifier) allows the searcher to link to the data/information at the data providers website. Once familiar with the data, the searcher can either be given the option of downloading data for further use, or if the data are linked to the data analysis and visualization toolkit, can apply the functionalities of the toolkit in a web-based environment.

Note that the EFIS was built as a cataloguing service, and not as a real Web service which would allow direct extraction of data/information from one or more data providers' websites. The system demonstrated that data, base maps and the system components can be hosted at different web locations. The challenges which resulted from the EFIS project were seen as input to follow up activities in building an operational system. The main points are listed below:

a) Evolve from a more fixed system architecture to one embracing new emerging standards in system interoperability
b) Further tailoring the system components both in terms of available tools and user friendliness
c) Further elaborate on the issues of metadata
d) Need for the system to be functional in the light of a growing amount of statistical and geo-referenced information
e) Introduce the concept of an EFIS to a larger audience of data providers and potential users

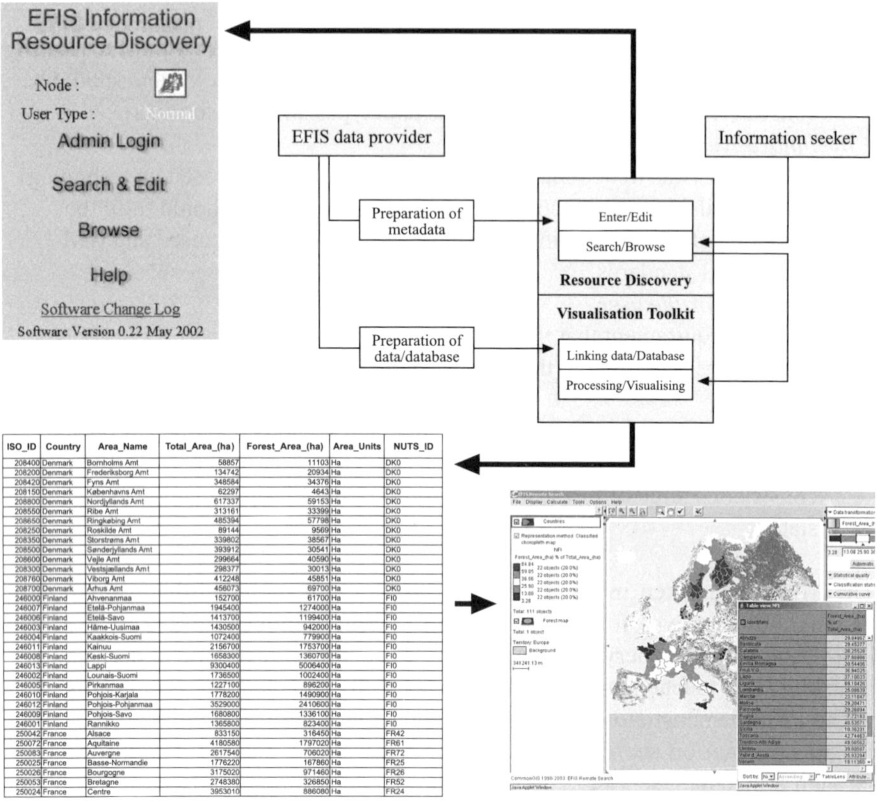

ISO_ID	Country	Area_Name	Total_Area_(ha)	Forest_Area_(ha)	Area_Units	NUTS_ID
208400	Denmark	Bornholms Amt	58857	11103	Ha	DK0
208200	Denmark	Frederiksborg Amt	134742	20934	Ha	DK0
208420	Denmark	Fyns Amt	348584	34376	Ha	DK0
208150	Denmark	Københavns Amt	62297	4643	Ha	DK0
208800	Denmark	Nordjyllands Amt	617337	59153	Ha	DK0
208550	Denmark	Ribe Amt	313161	33399	Ha	DK0
208650	Denmark	Ringkøbing Amt	485394	57798	Ha	DK0
208250	Denmark	Roskilde Amt	89144	9569	Ha	DK0
208350	Denmark	Storstrøms Amt	339802	38567	Ha	DK0
208500	Denmark	Sønderjyllands Amt	393912	30541	Ha	DK0
208600	Denmark	Vejle Amt	299664	40590	Ha	DK0
208300	Denmark	Vestsjællands Amt	298377	30013	Ha	DK0
208760	Denmark	Viborg Amt	412248	45851	Ha	DK0
208700	Denmark	Århus Amt	456073	69700	Ha	DK0
246000	Finland	Ahvenanmaa	152700	61700	Ha	FI0
246007	Finland	Etelä-Pohjanmaa	1945400	1274000	Ha	FI0
246006	Finland	Etelä-Savo	1413700	1199400	Ha	FI0
246003	Finland	Häme-Uusimaa	1430500	942000	Ha	FI0
246004	Finland	Kaakkois-Suomi	1072400	779900	Ha	FI0
246011	Finland	Kainuu	2156700	1753700	Ha	FI0
246008	Finland	Keski-Suomi	1658300	1360700	Ha	FI0
246013	Finland	Lappi	9300400	5006400	Ha	FI0
246002	Finland	Lounais-Suomi	1736500	1002400	Ha	FI0
246005	Finland	Pirkanmaa	1227100	896200	Ha	FI0
246010	Finland	Pohjois-Karjala	1778200	1490900	Ha	FI0
246012	Finland	Pohjois-Pohjanmaa	3529000	2410600	Ha	FI0
246009	Finland	Pohjois-Savo	1680800	1336100	Ha	FI0
246001	Finland	Rannikko	1365800	823400	Ha	FI0
250042	France	Alsace	833150	316450	Ha	FR42
250072	France	Aquitaine	4180580	1797020	Ha	FR61
250083	France	Auvergne	2617540	706020	Ha	FR72
250025	France	Basse-Normandie	1776220	167860	Ha	FR25
250026	France	Bourgogne	3175020	971460	Ha	FR26
250053	France	Bretagne	2748380	326850	Ha	FR52
250024	France	Centre	3953010	886080	Ha	FR24

Data table

Data analysis and visualization

Figure 2. Basic system architecture of EFIS (simplified).

f) Develop the system towards a reporting tool which can serve different user communities (in particular the European Commission) and processes (e.g. MCPFE, Kyoto)
g) Guarantee maintenance and further development over time

1.3.3 Network for a European Forest Information Service (NEFIS)

In order to operationalize a European forest information system at a broader scale, it will need to become more flexible in addressing: (1) diverse user needs, (2) data access policies and data rights, and (3) adequate and appropriate technological possibilities for the creation and presentation of value-added products (JRC, 2002a). The European Commission (EC) 5th Framework Programme Accompanying Measure project 'Network for a European Forest Information Service', or short NEFIS (contract number QLK5-CT-2002-30638) took the work initiated in EFIS further. NEFIS explored in more depth the issue of an overall information system architecture based on investigations of existing data reporting structures (i.e. from national level to EU or international level) in its 2.5 year running time. It gave explicit attention to the development of harmonized standards and procedures for providing data and metadata. The project partners representing data providers (and users) were made familiar with the concept of a European forest information system as NEFIS aimed to contribute to elaborating a framework for supporting potential data providers at various spatial scales (international, national, sub-national). The data providers were asked to make available datasets from their organizations and prepare metadata. The system developers in NEFIS developed an advanced version of the EFIS demonstrator. The datasets were then made available through the advanced demonstrator followed by an in-depth evaluation by all partners.

NEFIS, as did the EFIS project, strived to collaborate with ongoing information system development efforts in order to create synergies and guarantee an exchange of experiences. This was achieved by including the IUFRO, host of the Global Forest Information Service initiative (see section 1.3.5), as one of the project partners. Further contacts were established between NEFIS and the Canadian initiative of establishing a National Forest Information System (CCFM, 2004).

Due to the nature of the project (i.e. an Accompanying Measure) the expectations one may have from NEFIS should not be overestimated. However, NEFIS was very successful in advancing the understanding of such a system in relationship to new technology developments, its relation to existing reporting structures and processes and the role of metadata and metadata schemas, data rights and access policies. It also allowed numerous forestry organizations to be introduced to the philosophy behind a European forest information system, and gave exposure to the concept of metadata. Hence the project outputs can be considered as potential input for the building of an operational European forest information system in the foreseeable future.

In the following chapters of this research report we will highlight in more depth the individual outcomes of the NEFIS project including recommendations for future activities.

1.3.4 The European Forest Information and Communication Platform (EFICP)

Following the initiatives as described in the previous sections, the Joint Research Centre of the European Commission (Land Management Unit) in support of the DG Agriculture is coordinating and supervising the development of a European Forest Information and Communication Platform (EFICP) as part of it activities under its INFOREST Action (inforest.jrc.it/).

The EFICP is foreseen to contribute to improving the co-ordination, communication and co-operation between the Commission and the Member States, and between the Member States themselves, under the terms of Article 2(f) of the Council Resolution of 15 December 1998 on a Forestry Strategy for the European Union. It will give insight to the role and impact of the forest sector in the rural development policy. In particular, it will emphasize the contribution of the forestry measures to the fulfilment of: (i) the undertakings given by the European

Community and the Member States at an international level; and also (ii) the commitments made in the Ministerial Conferences on the Protection of the Forests in Europe, under the terms of Article 29.4 of Council Regulation (EC) No. 1257/1999 (EC, 1999). EFICP will contribute to enhancing the availability and quality of statistical and spatial information and facilitating the access of all interested parties to information about the forest sector.

The tender specifications (Invitation to tender n° 2005/S97-095822) state that the main aim of EFICP will be to:

- Serve as a forestry information and communication tool by providing, on a voluntary basis, a description in the form of metadata about available relevant information sources and datasets on the forest sector, including the forest measures within the rural development programmes under Council Regulation (EC) n° 1257/1999.
- Contribute to awareness-raising of the general public and practitioners as regards the implementation of EU Forestry Strategy by stimulating the information flow in disseminating available data and descriptive information, such as reports, analyzes and studies undertaken in the Member States on targeted subject areas.
- The system will be designed to allow users to locate and select data and information from different locations using the Internet and to visualize graphically available geographic and statistical information from these sources. The system will designed in a way so that whenever possible the data will remain with the data owner/provider. This will enable the data owners/ providers to maintain the data and provide access to the latest version of the data. The system will require that the data and information providers describe their assets with a reference metadata model in line with agreed standards.
- The system shall mainly rely on statistical, geographical and descriptive information that are linked to the forest sector and available at the level of the Member State as well at the international and the European Community level. The information that shall be made available by the Network must conform to the Community and national rules concerning the distribution of information, in particular in accordance with Council Regulation (EC) n° 322/97.

EFICP will therefore constitute a major step forward towards the definition and testing of means to efficiently retrieve and disseminate information on forests and other wooded land of the European continent, and facilitate communication and reporting on the forest sector. During system development specific issues will need to be addresses including: (a) the communication with other systems and initiatives, (b) the early involvement of relevant actors for system definition and characteristics to guarantee its acceptance, and (c) the building a flexible architecture based on open standards which will facilitate the maintenance and incorporation of new evolving technologies. These activities shall be linked to ongoing activities within the JRC (e.g. the Forest Focus data platform; The European Forest Fire Information System), the INSPIRE Geo-portal, projects such as NEFIS, information services (GFIS), research networks (the European national forest inventory network – ENFIN) and data collection agencies (e.g. Eurostat, UNECE/FAO) (Figure 3).

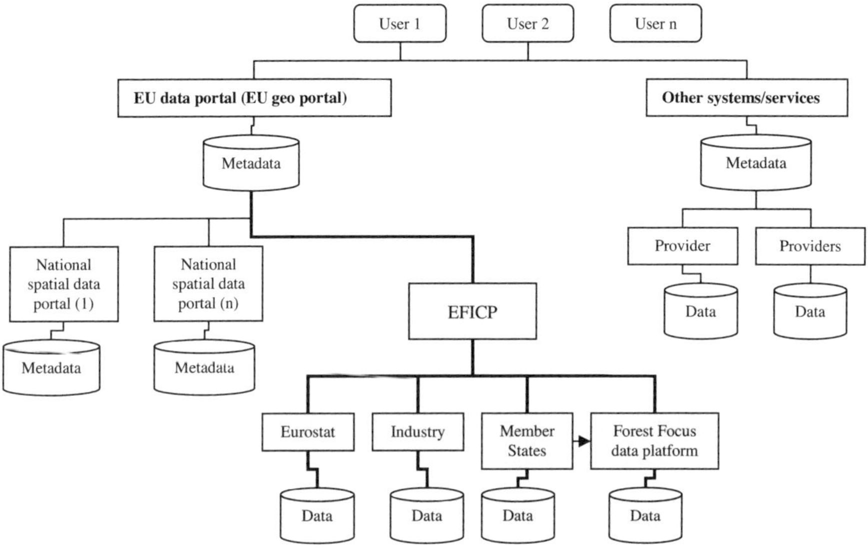

Figure 3. Possible application scenario for EFICP. Adapted from: Development of a European Forest Information and Communication Platform (EFICP). Tender specifications (Invitation to tender no. 2005/S97-095822).

The EFICP development will need to address in detail the following technical issues:

- interoperability
- access to data assets at both national and European/international level
- user registration and authentication/authorization in a federated network
- metadata reference model
- heterogeneous data retrieval, analysis and display
- identification of reporting tools

Many of the issues have been addressed in NEFIS and the project results can be of high value in the EFICP implementation phase during 2006 and 2007.

1.3.5 GFIS

GFIS is an initiative of the Collaborative Partnership on Forests (CPF). It is led by the International Union of Forest Research Organizations (IUFRO), together with the Food and Agriculture Organization of the UN (FAO), the Center for International Forest Research (CIFOR) and CAB International. A range of additional partners contribute information to GFIS.

The role of GFIS is to add sufficient value to currently available information such that GFIS becomes the global gateway of choice for a critical mass of partners providing and using information. The vision of GFIS is to bring together all stakeholders and users to communicate their on-line forest related information resources through the GFIS gateway.

The mission of GFIS is to enhance access to all types of forest information for all partners, including governments, researchers, forest managers, NGOs, community groups and the public at large; and to contribute to an improved understanding of complex forest-related issues, to enable better decision-making and to facilitate a more informed public engagement in forest policy and forest management at all levels.

The partnership development of GFIS has been realized as an internet gateway that provides access to forest-related information through a single entry point. All information available through GFIS is provided by partners around the world concerned with forest information. Partners also contribute to system and partnership enhancements, as well as capacity building efforts to strengthen developing country institutions.

GFIS resource discovery is based on a metadata approach that focuses on key elements of data and provides information to different type of information resources and therefore facilitates overall discovery of information. GFIS provides an open exchange standard for its information categories. The standard helps partners to generate their inputs and allows them to manage their contributions to GFIS. The standard is based on the Dublin Core Metadata Initiative (DMCI) and on AGRIS (FAO) metadata schemas. GFIS provides full documentation of the information exchange standard, as well as a control panel where partners manage their contribution details through the gateway.

The NEFIS project (and before that the development of the EFIS demonstrator – see section 1.3.2) was implemented keeping close contact to the ongoing development of GFIS (GFIS, 2005).

2. The Overall Architecture of a Forest Information System

Following the introduction on the state-of-the-art of a European forest information system it seems appropriate to start with a generic view on the architecture of such a system. This chapter will introduce such an architecture, taking into account a set of examples of existing reporting processes and data flows which exist at a European level. These examples or "use cases" were seen as complementary information in order to design the overall system architecture. Chapter 2 is divided into four parts: 2.1 "The Unified Modelling Language; 2.2 The user view – overview of European forest information processes and their principal actors; 2.3 "Commonality between use cases – the Big Picture " (a study of existing use cases); and 2.4 "NEFIS architecture". These were set out as tasks in the NEFIS project.

2.1 The Unified Modelling Language (UML)

The Unified Modelling Language (UML) has been adopted by the NEFIS project as a means of structuring discussion and formulating the information systems requirements and software specification for development of a system that might satisfy pan-EU needs for an integrated European forest information system (EC, 1997; Päivinen and Köhl, 2005).

UML has been developed and maintained by the OMG (The Object Modelling Group) a not-for-profit consortium of virtually every large company in the computer industry plus hundreds of smaller ones whose aim is to maintain computer industry specifications for interoperable enterprise applications. It became the software engineering de-facto gold standard as an extensible graphical language for visualizing, specifying, constructing and documenting the artefacts of a software-intensive system. As a language it has a rich syntax and formal semantics which include mechanisms for extending the vocabulary and grammar to encompass and express specific needs of any application domain of interest. UML is now an accepted ISO specification (ISO/IEC 19501). It is employed not only for writing the systems blueprints, application structure, behaviour, and concrete architecture things such as programming language statements, database schemas, and reusable software components, but also for describing and modelling conceptual aspects such as business processes and system functions (OMG, 2005b) .

It has been recognized that the real value of the UML to NEFIS is to support the elicitation and analysis of requirements and the identification and modelling of the information system architectural framework. Further UML was seen as a vehicle for system developers and forestry domain experts to establish a common framework of understanding on an overall architectural design. This was done by transcibing textual use cases provided by forest experts into UML design.

2.2 The user view – overview of European forest information processes and their principal actors

2.2.1 Introduction

Information systems are perceived by different actors (stakeholders) in radically different ways. System users may have little idea of the way in which a system functions behind the user interface, nor should they care – so long as the system delivers their requirements in a correct, meaningful and consistent manner. As we saw in the introduction it is the user's requirements that govern the system's functions, in turn the system's functions determine how it is implemented. It is therefore fundamental to understand the user's perspective, or Use Case View, in order to specify a system that can satisfy the user's requirements. Figure 6 in section 2.3.3 illustrates the way in which the User View is overarching in relation to the Design, Process, Implementation and Deployment perspectives.

The ways in which forest information is measured, collected, utilized and presented at the European and international levels are multiple and various. There are many actors, stakeholders and users of the information throughout the forest and related sectors. The various international conventions and agreements that are concerned, to a greater or lesser extent, with forests each have their own objectives and reporting requirements, although efforts to standardize reporting are ongoing. Furthermore, the nation states that make up the European Union each have their own forest sector with its own historical and political contexts. The picture of forest information at the European level is therefore a complex one, and one in which there are many overlapping interests and requirements. Furthermore, these various forest information paradigms operate at multiple levels – global, EU, national, sub-national and local – further complicating things.

The approach within the NEFIS project has been to compile a core set of characteristic European forest use cases. These use cases have been developed as brief, textual descriptions of selected information processes. Where appropriate UML use case diagrams were produced for each of the information processes. The purpose of the use case diagrams is to provide a better understanding of how a European forest information system could benefit the forest community by fulfilling specific information communication and processing tasks in addition to identifying information resources.The following use cases were prepared within the NEFIS project.

- UN-ECE/FAO Global Forest Resource Assessment (FRA) and Temperate and Boreal Forest Resource Assessment (TBFRA) 2000 (Regional FRA)
- MCPFE Criteria & Indicators for Sustainable Forest Management
- Forest Products Production and Trade Flows
- Land Use and Land Cover Change for UNFCCC

In developing these use cases it became clear that some of these information processes are very complex and their full development was beyond the scope and mandate of the NEFIS project. The UML exercise carried out within the project was thus concerned with presenting an overview of a number of central information processes rather than the minutiae of individual processes. Therefore, developing the use cases to a very fine level of detail would not serve the project's objectives.

Section 2.3 of this report deals with the commonality between these use cases and takes those as foundations for building a generic framework/architecture for a European forest information system.

Here we will briefly consider UML Use Case Diagrams. A Use Case Diagram represents a user's view on the system and is comprised of:

- Actors – people, or other systems, that interact with the system
- Use cases – the normal behaviour of an entity
- The relationships between actors and use cases. Relations can be of the following types:
  - Communicates – an association between actors and use cases.
  - Extends – an association between use cases where a base use case may, or may not, include the behaviour of the extending use case.
  - Uses – an association between use cases where a base use case includes the behaviour of an extending use case.

A simple Use Case Diagram is shown in Figure 4. A Forester [actor] may want to measure 'diameter at breast height' (DBH) distribution [use case] of a stand of trees. To do this she has to measure DBH [use case] of a sample of individual trees. The base use case (Measure DBH distribution) uses the behaviour of the common use case (Measure DBH).

An example of a Use Case Diagram is Figure 5 for the Temperate and Boreal Forest Resource Assessment. Note that a Use Case Diagram may contain many individual use cases. Each use case has a specified goal and a set of inputs and outputs.

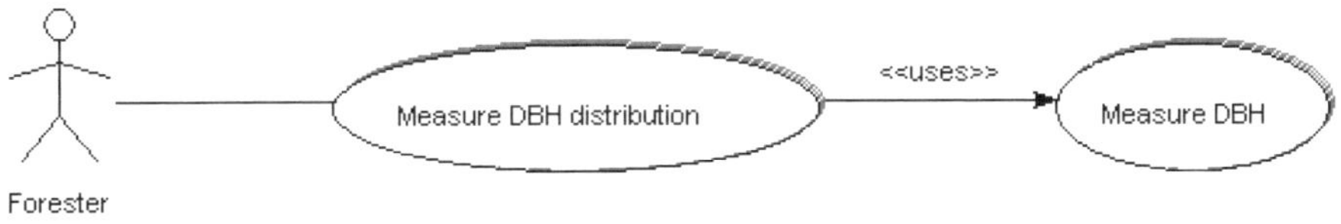

Figure 4. Example Use Case Diagram including Actor, Use Case, Associations.

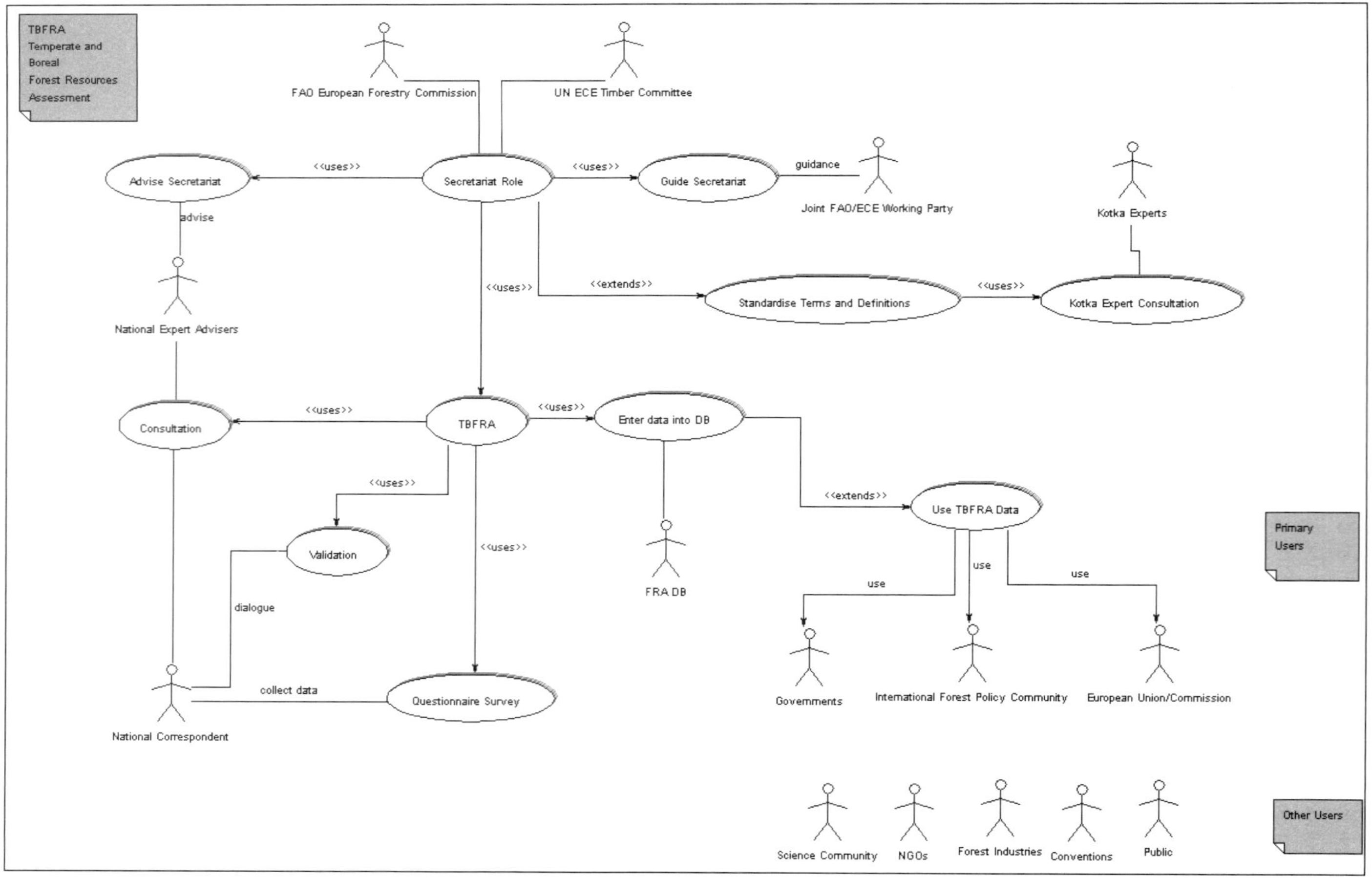

Figure 5. Example Use Case Diagram for TBFRA. For further details of the use case diagrams and textual description, readers are referred to the NEFIS UML model (deliverable 9 of the NEFIS project which is in the NEFIS Knowledge Base http://www.nefis-kb-info).

2.3 Commonality between use cases

2.3.1 Introduction

In this section we briefly review information systems engineering issues that affect the design and implementation of a European forest information system. Generic use cases are identified and their significance for the system requirements, technical options and architectural framework are discussed. UML is then used to model the high level relationships and information processes recommended for incorporation into a broader model of forest information processes at the pan-European and global levels.

2.3.2 System design issues

The scale of forest information is not only vast – ranging from remote sensed images to tabular summaries of regional statistics – but also constantly evolving with time and distance scales. They can range from seasonal growth of individual trees and stands to forest change at only two or three cutting cycles per century.

Much of the discussion within the forest information systems community has thus far focussed on the 'functionality' and 'usability' of a future system. Functionality being defined by the ISO 9126 standard for software quality characteristics (ISO, 1991). They include suitability, accuracy, interoperability, compliance and security; and Usability being viewed as the system's understandability, learnability and operability. This discussion led to a number of questions raised by the earlier projects dealing with the development of a Euroean Forest Information System (EC, 1997; JRC, 2002b; Päivinen and Köhl, 2005): (1) What information is processed? (2) Where is it maintained? (3) How are the resources shared and distributed? and (4) Where and how is this information processed? Therefore, as part of the NEFIS project, example systems have been further developed to prototype the use of metadata for resource discovery and to demonstrate geographic information system-(GIS-)based user interfaces for forest information visualization and exploratory analysis.

There are a number of factors that will affect the architectural infrastructure of a forest information system: (1) the systems should exploit advanced spatio-temporal data collection and information management; (2) they must entail dissemination and fusion of heterogeneous distributed information; (3) and they require sophisticated analysis, modelling and visualization of information which will outlive the diverse hardware and software platforms on which they reside. Therefore, certain features of the systems become important. From a user perspective these are its efficiency in terms of time and resources, and its reliability – i.e. fault tolerance and recoverability. However, from a systems design and implementation view point the critical factors become: (a) portability (as defined by the system's adaptability, replaceability, conformance, and installability); and (b) its maintainability (in other words: analyzability, changeability, scalability, stability and testability), because of the evolving nature of the information and applications.

It should be noted that a rule of thumb used in the information systems community suggests that 80% of the cost of a software system will be in the post-deployment maintenance phase (e.g. see (Endres and Rombach, 2003) the bulk of which will be adaptive and perfective maintenance. Maintenance refers to the response to changes in the environment such as operating system upgrades and implementing additional requirements such as form and report generation. It is these latter factors which have been major concerns and have directed our recommendations toward compartmentalizing any technical design solution with component and service oriented federated technologies as described in the following sections.

2.3.3 Analysis and design premises

The goals of a European forest information system are common to most large-scale systems in that the system must aim to:

- improve efficiency and reduce duplication of effort;
- reduce the redundancy and duplication of data; and
- add value and confidence to information, by facilitating validation and single sourcing of information.

Furthermore, because of the changing nature of the information and its distribution across many countries and regions, the information system must address problems of semantic interoperability and support version control. Discussion with potential users suggests the proposed system should have a minimum impact on current business practices, and must support and not constrain users.

Numerous studies of large-scale commercial and governmental information systems have shown that the bulk of failures are not developmental or technical, but operational. Once information systems are delivered and working they: (a) are simply not used, (b) used briefly but then abandoned, or (c) their value to the enterprise does not justify the cost of development and ownership (Glass, 2003). The real issue therefore becomes: *Why would someone want to use this system?* The history of information systems has shown that if the answer is "because they are obliged to", they will subvert or neglect the system processes so the data becomes outdated and worthless. If the resource user has to pull down menus, click buttons and boxes, cut and copy fields many tens of times to fulfil their role and glean the information for a report they will cut corners. Similarly if the information provider has to fill in tens of metadata fields in a form before submitting their data when previously they simply e-mailed a spreadsheet to, for example, a contact person at a data collection agency they are liable to get careless and make mistakes.

This problem has led to the contemporary software development process focussing on the identification of the end-users of a system and meeting their needs, expectations and requirements. This approach will ensure that the system provides direct benefit to its target end-users. Indeed it is now considered best current practice to actively involve the users at every stage in the development process (Pressman, 2005). This starts with what is termed 'use case' analysis, a process of identifying the users ('actors') and the use they would make of a system to gain

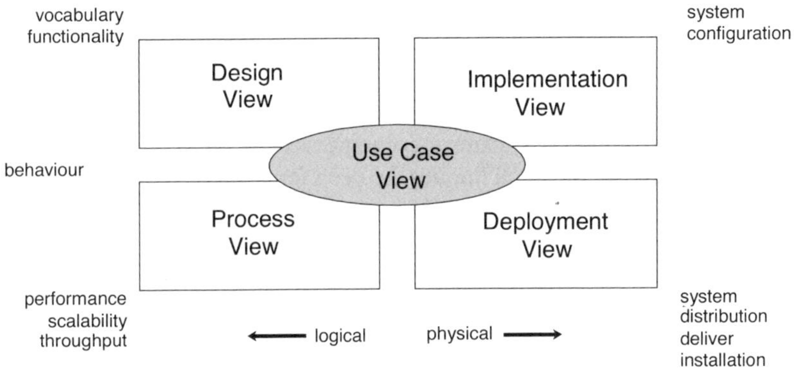

Figure 6. The Use Case View (Kruchten, 1995).

benefit from it ('use cases'). These use cases are then refined to not only identify the functionality and nature of the required human–computer interaction, but also extract the system objects, their interrelations and their deployment (see section 2.2.1). The use cases thus form the core which bridges the various views from logical design through to technical architecture and implementation (Figure 6).

2.3.4 User needs

There are clearly a huge range of actual and potential users and applications, and it is not feasible using current requirements engineering techniques to map even a moderate proportion of the potential use-cases. However by exploiting the work of Kammersgaard on Human–Computer Interaction (Kammersgaard, 1990) it is possible to partition the user group into four classes (Figure 7) by distinguishing whether the focus is on individual use or within a collective context, and whether the primary design principle source is from a contents or expression level.

Use content / Source of design principle	**Collective**	**Individual**
Expression Level	Systems perspective	Dialogue Partner
Contents level	Media perspective	Tool perspective

Figure 7. Perspectives of Human–Computer Interaction (Kammersgaard, 1990).

The four perspectives of NEFIS actors can therefore be viewed as:

1. *Systems Perspective:* in which the user plays a small part in the overall process. The classic example of this type of user is the check-out operator in a supermarket. In forest systems it encompasses raw forest information data entry and form filling and implies the need for a classic transaction processing architecture.
2. *Media Perspective:* here the system is an active media for communication, for example publication of press releases, educational resources and computer-based conferencing. The actor needs may be satisfied by multi-media technologies delivered via web portals.
3. *Tool Perspective:* in the tool perspective the system provides sets of tools and facilities for expert use, e.g. raw datasets, model archives, statistical analysis and visualization tools to support the forest researcher.
4. *Dialogue Partner:* here the system and user are regarded as partners in a dialogue communicating to get some task performed. This implies classic decision support system architecture and often requires knowledge base support.

It is suggested that the majority of clearly identifiable NEFIS users fall into the dialogue partner category. The quickest initial significant benefits may be gained by addressing the needs of this group of user. In the shorter term the bulk of needs of the other groups may be satisfied by web portal and gateway services. A number of specific use cases from the dialogue partner category were therefore identified and documented (see list in section 2.2.1).

Commonalities between the textual descriptions of the use cases were identified in order to build generic use cases and identify the boundaries of the system. These are presented in Figure 8.

2.3.5 NEFIS generic use cases

The common patterns that arise when analyzing each of the dialogue partner applications form a set of processes and functions. Thus depending on the user role and task, they all work to a set of guidelines, conventions, standards or directives which specify the required information elements to be used for and generated by the task (e.g. MCPFE criteria and indicators). For each of these information elements the user must locate a source, request and retrieve the dataset which must then be validated. At this point it may become necessary to transform the data into a particular format, or allow for the harmonization of the retrieved data. This step may be compared to the classic data-mining ETL (Extract, Transform, Load) process. Having retrieved and 'normalized' the required data, they are then collated and aggregated into a dataset. The dataset is then transformed through an aggregate model, for example statistical analysis. The aggregate must also be formatted for publication, for example into tabular or graphical form, before it is made available to others via a publication process. That process could range from writing to a public file, document, a database or even web page. Not only may the whole process

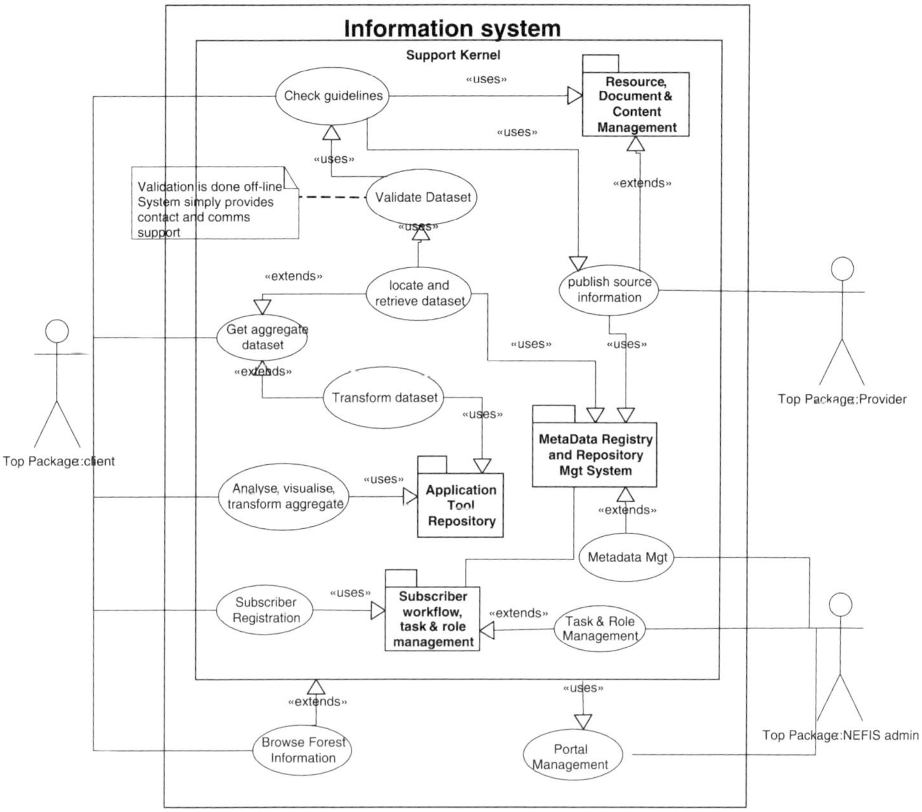

Figure 8. Generic dialogue partner process for a European forest information system.

be repeated, but each step may be iterated often tens, and in some cases hundreds, of times. Finally it should be noted that the published result or output from one task becomes input dataset for another. Thus elements from a national inventory may form the input to forest resources questionnaire (as for example prepared by the UNECE/FAO). The questionnaire results and their data may then feed into their Regional Forest Resources Assessment Report, which provides input to the MCPFE report on the State of Europe's Forests, or the Global Forest Resources Assessment publication of the FAO. As part of the collation, aggregation and publication steps, care must be taken to ensure the transparency of the origins of data and the processing steps so that: (1) contributors are appropriately acknowledged; and (2) the end-users have sufficient information to evaluate whether the data are appropriate for the intended use.

2.4 NEFIS architecture

2.4.1 Core subsystems

The generic use case analysis exposes functionality which must be supported in all of the dialogue partner applications and from this the necessary subsystems can be identified. These core subsystems can be grouped into four packages:

1. resource and content management;
2. tool and component repository;
3. task and role support; and
4. metadata management.

Because of the huge and disparate range of resources the first of these, *Resource and content management*, becomes essential not only for asset management, but to support the classic change and version control and configuration management demanded by the evolving nature of the resources. Content management services also provide the infrastructure for resource description, discovery and integration, and therefore control content assets used to dynamically create and deliver web presence. Coupled to resource management is the need to maintain a *Tool repository* of both the domain specific and task support tools for users. It is particularily important for resource publication, standardization and harmonization, and also standard components for developers adding to the system. They can be API (Application Program Interface) libraries, parsers for XML-based markup languages (e.g. GML the Geographic Markup Language, or an FML – a possible forest markup language), metadata selectors and architecture adaptors. The tool repository could also store task oriented common forestry GIS, visualization, analysis and reporting tools and provide the infrastructure for a forest model archive. Provision of *Task and role support* enables not only subscriber access, and authorization control and data rights management. It may also dramatically improve the usability of the system by forming the infrastructure for: (a) identification of and navigation through datasets of interest; and (b) for standardization of documents and form templates. In turn this potentiates workflow management, choreography and automation of repeated processes.

The discussion of metadata within the NEFIS project has rightly focussed on forest themes and terms. However, it can be seen that by careful partitioning, the NEFIS can be constructed iteratively and extensibly from independent communicating subsystems. To ensure the exploitation of highly cohesive, lowly coupled components within what is essentially a distributed heterogeneous system requires a common interchange language and middleware with agreed terminology – see for example the CORBA initiative (OMG, 2005a). Therefore, metadata is not just needed for resource discovery or informed decision support, but it is also the glue that ensures the semantic interoperability (agent dialogue) of the system. *Metadata management* facilities therefore form, what may be argued to be, the central component of the NEFIS subsystems.

2.4.2 Identifying the NEFIS architecture

The architecture of an information system may be defined as a framework that describes its form and structure i.e. its components and how they fit together. The architecture may be viewed either in its logical form (as the relationship between the conceptual elements of the system) or in its physical form (the components and subsystems reside in the implementation deployment).

A good logical architecture should ensure separation of concerns. For example at the most basic level, the storage of persistent data (i.e. databases, files, etc.) should be managed separately from the application processing (such as statistical analysis). Application processing should itself be distinct from user interface or presentation concerns. Those would be for example web pages or visualization tools. An example of this logical layering is given in the European Computer Manufacturers Association reference architecture (NIST, 1993) presented in Figure 9.

This decoupling not only facilitates reuse and incremental development, but, when specified with well defined interface protocols, can address the interoperability, maintainability and scalability concerns discussed earlier. Figure 10 (from Melnik and Decker, 2000) shows the role metadata plays not only in ensuring interoperability between heterogeneous application nodes, but also interoperability between architectural layers, and therefore future-proofing and bolstering extensibility.

Because of technical concerns such as security, communications bottlenecks and device overloading, and political issues such as resource ownership and information rights management, any logical architecture may ultimately be compromised by its physical implementation. This issue was raised in the report by (Päivinen and Köhl, 2005) without conclusion. A number of options exist for a potential physical architecture. Conceptually the simplest being a 'Repository Model' with shared data held in a central database and subsystems maintaining their own information. This model has clear advantages in that it addresses the problems of information redundancy and duplication. It therefore guarantees consistency and confidence. There are, however, numerous reasons why this would not be feasible for a pan-European system. They include, for example, the scale, range and distribution of

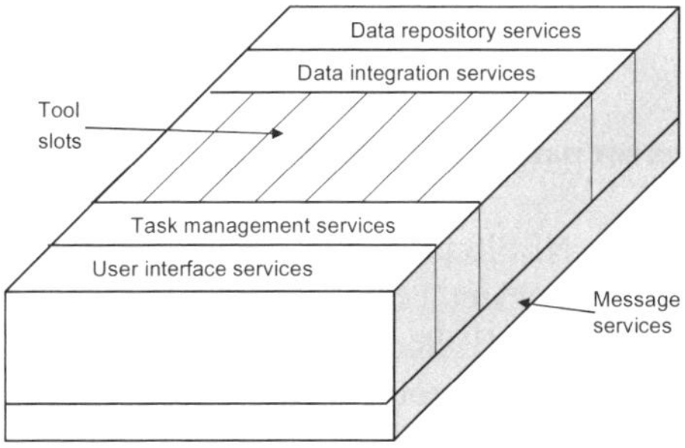

Figure 9. European Computer Manufacturers Association reference architecture (NIST, 1993).

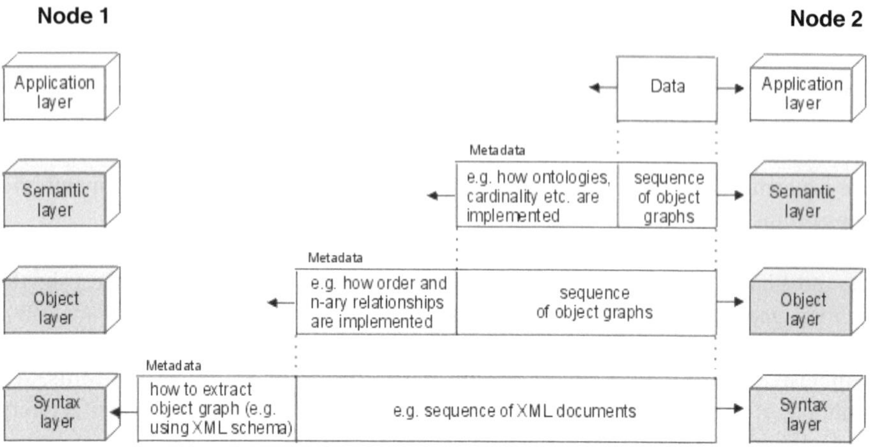

Figure 10. Metadata, Architectural Interoperability and Protocols (Melnik and Decker, 2000).

forest information. As this information can, to a large extent, be compartmentalized a viable alternative to the central repository model would be the 'Client-Server Model'. Here a set of networked servers offer services to other sub-systems and clients call upon the services offered by these servers. The clients may be classed as 'thin clients' or 'fat clients'. For thin clients application processing is carried out on the server, and the client is only responsible for the presentation layer. In the case of fat clients the server is only responsible for the data management layer. The application processing and presentation layers reside on the client node. Client-Server systems address many of the technical problems that a large-scale forest information system presents. The heterogeneous nature of the distributed subsystems however may be at the cost of coherence. Contemporary 'Distributed Object Architectures' address these problems by building upon the server concept but by removing the distinction between client (receiver of services) and server (provider of services). The architecture is based on system components promoted to first class objects that provide an interface to the set of services they provide. These objects are distributed on a network and communicate through a standardized middleware compliant request broker.

2.4.4 Web services for pan-European forest information

Although novel when the NEFIS project was being proposed, there has been a rapid development of distributed object technologies. They exploit the power of the world wide web through use of XML/SOAP as an RMI (Remote Method Invocation) message based interoperability protocol supported by the Universal Description, Discovery and Integration (UDDI) directory mechanism. These 'Web service' technologies were identified and discussed at length in a technical review within the NEFIS project (Rennolls et al., 2003). The support for metadata based resource discovery provides an elegant solution to problems of heterogeneous distributed

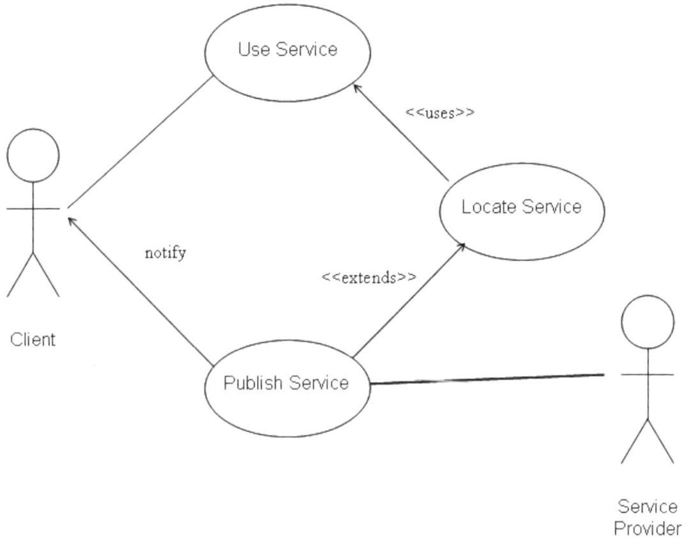

Figure 11. The Publish-Subscribe Pattern.

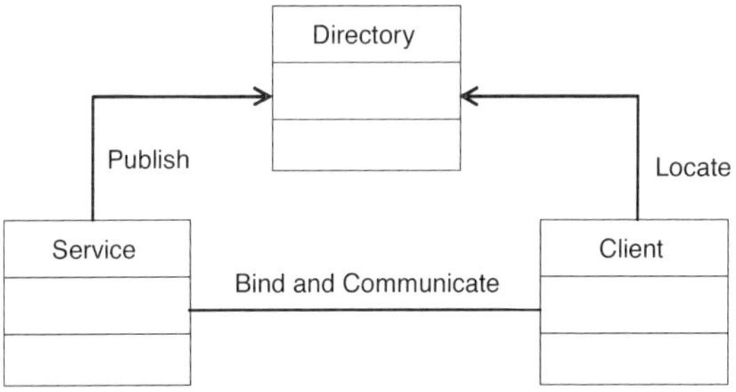

Figure 12. The Service Oriented Architecture.

application, integration and interoperability. This includes the problems arising from maintenance and integration of legacy systems, and has ensured that these technologies have since become the key enabler of large-scale enterprise systems and form the basis for the 'semantic web'. A simplification of the generic European forest information system use case reveals the 'Publish-Subscribe' pattern (Figure 11) of this 'Service Oriented Architecture' (SOA).

As suggested, the web-enabled SOA ideally addresses concerns of business process integration, workflow and system transparency (access, location, scalability, etc.) found in the EFICS and EFIS feasibility analysis (JRC, 2002b; Päivinen and Köhl, 2005). Conceptually the structure is simple (Figure 12). It is based on collaborations between a 'publisher' of a service (Figure 13) and a 'subscriber' who locates and uses those services (Figure 14) via a directory service (Figure 15).

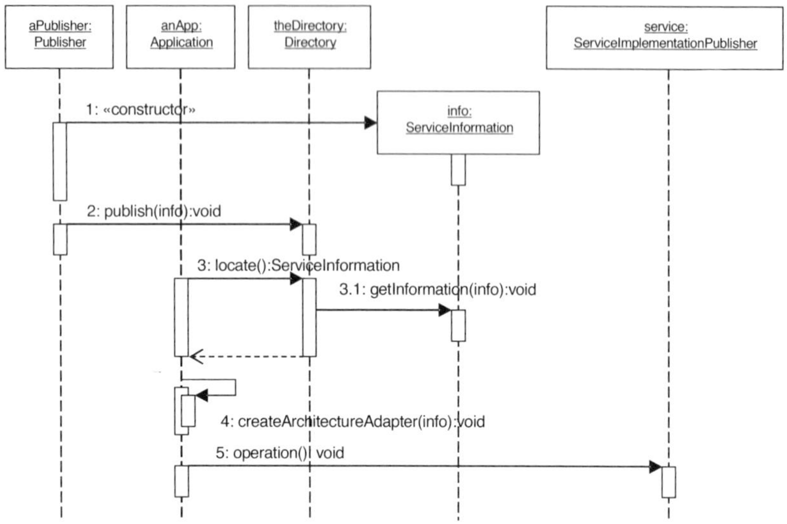

Figure 13. SOA Publication Interaction.

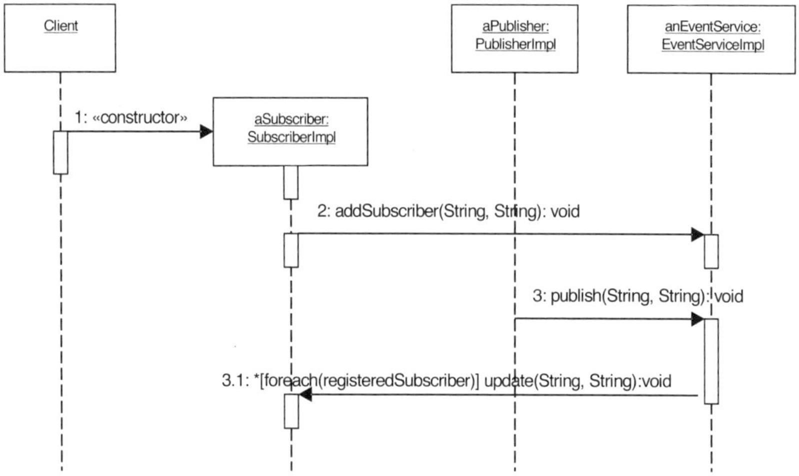

Figure 14. SOA Client Registration.

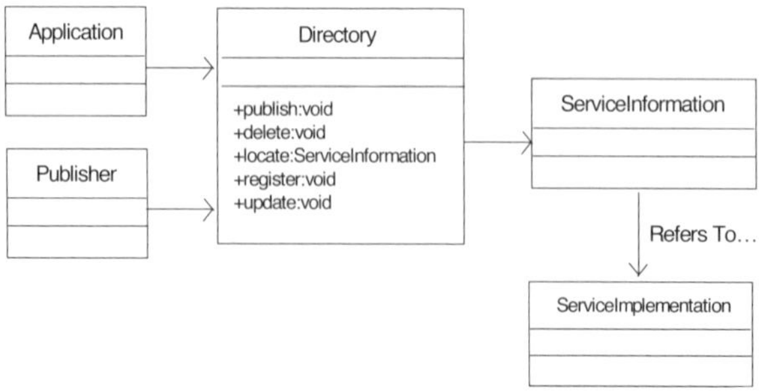

Figure 15. Service Directory Class Diagram.

This technology is not proprietary, but is instead supported by huge repositories of pluggable components and open-source code adhering to open system standards. Such is backed by nearly all of the largest computing and software companies. Including the core design elements of the Web such as identification of resources, representation of resource state, and protocol-based interoperability that support the interaction between agents and resources potentiate federated (distributed) data warehousing. Current work led by the World-Wide-Web Consortium (W3C) is converging standards for SOA Web services (W3C, 2005) with those for Grid computing. A computational grid is defined as "a hardware and software infrastructure that provides dependable, consistent, pervasive, and inexpensive access to high-end computational capabilities" (Foster and Kesselman, 2005).

Essentially Grids coordinate resources that are not subject to centralized control, and use standard, open, general-purpose protocols and interfaces, and deliver non-trivial qualities of service (GlobusAlliance, 2005). It can therefore be seen that the Web service oriented architecture presents the best option for structuring future European forest information systems.

2.4.5 NEFIS deployment

Although the SOA appears to address all of the functional, technical and political issues identified earlier, one key problem remains for a European forest information system (and other multi-jurisdictional projects of a similar nature). That is, huge volumes of data currently reside at the national level. An example are national forest inventory data. Unfortunately there are no universally applied procedures for the collection of these data in terms of methodology and data storage. To address the different uses and products of their respective forest resources, different countries have historically employed different definitions of the base data elements. Examples are tree top-height (in some cases measuring to the highest tip, in others measuring to the point where the top stem has a 3, 7 or 7.5 cm diameter) or forest area (with different countries measuring wooded areas of varying minimum size and with anything from 30% to 100% crown cover). In order to provide 'objective comparable data' at a supra-national level, data must undergo a certain degree of harmonization. This problem will always exist, even if pan-national standards were introduced, as research and decision-making would still need to be based on trends which must include this historical, legacy data (EC, 1997; COST E43, 2004; Köhl et al., 2000; Päivinen and Köhl, 2005). Although simple linear conversion functions are often employed, some aspects of harmonization potentially involve complex, non-trivial models. It can be seen that with the large volumes of un-normalized data originating from distributed servers, standardization or harmonization on each query will result in a much degraded performance of any distributed information system.

A similar problem to harmonization, that is common to any large-scale database system, and may be viewed as part of the validation process discussed earlier, is that of 'data-cleaning'. It is clear that the harmonization and data-cleaning should only be performed once, but this raises the question of where this cleaned data should be held? There are strong arguments (see for example Rennolls, 2005) for storing and maintaining this data in a controlled archive repository on a central server. A number of technical and social factors mitigate against this. Technically the largest problems stem from server and network bottlenecks; socially, ownership issues arise. However, Web service technology now supports federated (distributed) warehousing, and so there is no technical reason why the information should be held centrally and not be distributed. This is linked to the principle that the producer/publisher of data should maintain control and ownership. As was discussed earlier, one of the features of the NEFIS requirements is that one person's output is the next person's input. Similarly, one person's metadata is another's data – i.e. each node must support and act as both subscriber and publisher. It is therefore recommended that information is held where it is processed: e.g. raw data is held in forest management units or national inventories; harmonized data published by country correspondents should be held at the national level; data which has been cleaned, validated and standardized as part of an aggregation process should be maintained, published by, and held on the servers of the agency that performs the process. At a European level it is reasonable to assume the national and pan-national agencies have the server technology in place or can add the pre-requisite components to existing forest information systems to support this. However, it should be noted that if the principle is extended to a global level, centralized hosting services may have to be provided for countries with no existing forest information systems. A summary of an architecture which addresses the issues discussed thus far is presented in Figure 16. In essence this is a distributed service-oriented architecture in which the publisher maintains ownership and control of published information at their own node. Visualization, analysis and other processing of information are done at the fat client subscriber node.[3] Once processed, roles are switched and the subscriber becomes publisher. Finally the 'NEFIS server' exploits Web services interoperability technologies to provide the glue service discovery, gateway services and acts as a central tool and resource repository.

[3] This is one approach. The [Canadian] National Forest Information System takes a different approach and is based on a thin client subscriber node with heavy server side processing.

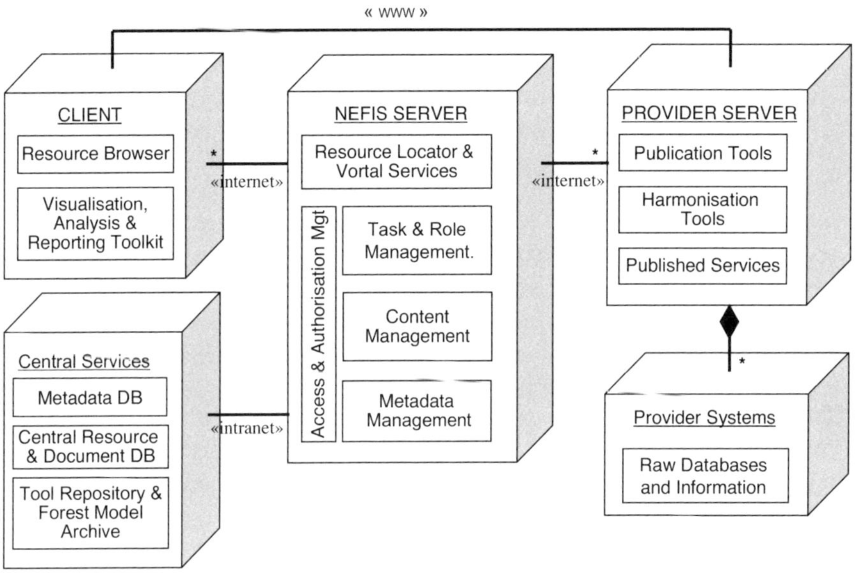

Figure 16. NEFIS Deployment.

3. Development of a Metadata Schema

3.1 Metadata

The extraction and analysis of data from a variety of sources in order to draw conclusions for decision-making, when carried out by a human data analyst, involves a number of stages including:

- Identification of appropriate sources
- Evaluation of those sources for relevance and reliability
- Extraction of required data
- Manipulation using tools appropriate for the required purpose
- Interpretation of the compiled results
- Presentation to the appropriate audience(s)

At each stage the analyst will make use of ancillary information regarding the datasets selected, the requirements of the analytical tools to be used, the expectations of the intended audience and their familiarity with the subject matter. Throughout the process, the analyst's actions may be modified after discussion with others, and the final conclusions are likely to be subjected to some kind of peer review, whether through a publication process or internal assessment by those who commissioned the work.

This knowledge generation cycle of retrieval, analysis, publication and storage remains fundamentally unchanged by the widespread use of Information and Communication Technologies (ICTs), but the mechanisms, scale, accessibility and audience reach have become very different. Data gathered for a specific purpose for a limited audience can potentially be retrieved and used for entirely different purposes or audiences. This carries both benefits and risks: consolidation of incompatible data could lead to erroneous conclusions with unpredictable results. It is essential that the users of automated systems are aware of the degree of risk entailed at each stage in the process, and thus emphasis needs to be given to clear and transparent description of resources including comprehensive information on their quality, resolution, scale of capture, currency of information, authoritative source, etc. (see section 3.4.3 on quality issues). This will require that the background information a human analyst may draw on (often 'in his/her head') should be explicitly stored with the data, and be retrievable as needed throughout the knowledge generation process. That is the central purpose of metadata.

Therefore, one of the tasks in the NEFIS project, was to employ metadata to enhance both a clear description of information resources and address interoperability in a standards-led process that has long-term sustainability.

The US E-Government Act 2002 defines 'interoperability' as:

> *...the ability of different operating and software systems, applications, and services to communicate and exchange data in an accurate, effective, and consistent manner.*

The challenge facing information providers, governments, and the EU is to ensure that the data/information/knowledge produced can be found and accessed (easily) by its target audience. In the EU the IDA programme (Interchange of Data between Administrations) has played a pivotal role in addressing the issue of interoperability. The 2003 working paper *Linking up Europe: the Importance of Interoperability for eGovernment Services* (IDA, 2003) clearly articulates the case for interoperability not just within administrations, but also with the business sector and the public, seeing it as fundamental to the success of the European enterprise. It stresses the need for:

> *...information and services that are developed from a 'customer-centric' viewpoint while recognizing that the reality today is the emergence of 'islands' of eGovernment that are frequently unable to interoperate due to fragmentation resulting from uncoordinated efforts in developing the services, at all levels of public administration.*

The EU's plans for addressing customer centric services are set out in the *eEurope 2005 Action Plan* (EC, 2002). Its objectives relate to the main areas of e-government, e-health, e-learning and e-business rather than 'e-research'. Nonetheless, the infrastructure and standards set in those areas will impact strongly on the IT developments initiated in the EFIS and NEFIS projects, and continued by their successors, in particular the European Forest Information and Communication Platform (EFICP). As the IDA paper notes:

> *Traditionally, organisations have developed hierarchical organisational structures to serve well-defined user communities, each with their unique ways of processing information. This hierarchical framework has resulted in closed, vertical, unscaleable and frequently proprietary information systems that mimic their paper-based predecessors and cannot share information across internal structures, let alone with other organisations (IDA, 2003).*

This is certainly true in the forest sector and it must be a primary objective of the successors to EFIS and NEFIS to remedy that situation.

Providing good access to information, however, does not automatically increase its use. In a review of progress on the eEurope Action Plan the Commission notes:

> *...from the evidence available, it seems that progress in e-government supply is not matched by proportional increase in demand. To address this problem, e-government policies across Europe are shifting towards user-oriented and "one-stop-shop" approaches, e.g. presenting services on the basis of "life-events" and "business episodes". But greater efforts are needed, notably on public awareness, usability, multi-channel access, and user confidence in relation to privacy and security issues, including identity management (EC, 2004a).*

In terms of a European forest information system confidence in the 'brand' must be built by ensuring that the data access on offer corresponds with user (government, industry and public) needs. The cost of providing it on an ongoing, sustainable basis will be considerable; however, deciding not to offer access may not be an option if the producing agency is required to do so under e-government regulations in the future.

The European Commission published a European Interoperability Framework (EIF) in 2004 (EC, 2004b). The EIF will act as a reference document on interoperability for the IDA successor program IDABC (Interoperable Delivery of pan-European eGovernment Services to Public Administrations, Businesses and Citizens) which will run from 2005–2009. A second document under preparation by the EC will address long-term implementation and maintenance of the EIF. These documents are likely to have significant influence.

The EIF defines an interoperability framework as:

> *...a set of standards and guidelines which describe the way in which organisations have agreed, or should agree, to interact with each other.*

The three aspects of interoperability considered in the EIF include:

- Technical interoperability: the technical issues of linking up computer systems, the definition of open interfaces, data formats and protocols, including telecommunications;
- Semantic interoperability: ensuring that the precise meaning of exchanged information is understandable by any other application not initially developed for this purpose; and
- Organizational interoperability: modelling business processes, aligning information architectures with organizational goals and helping business processes to co-operate.

The Computing Technology Industry Association (CompTIA) believes that technical interoperability is close to being achieved through open standards (CompTIA, 2004). It does not wish to see any particular technology mandated by the EU, believing that existing fora such as the World Wide Web Consortium (W3C), Organization for the Advancement of Structured Information Standards (OASIS), and Web Services-Interoperability Organization (WS-I) have demonstrated their ability to define standards which are 'best-of-breed' and the industry is happy to adopt. It sees semantic and organizational interoperability as much more challenging, and calls for a strong industry focus on these areas.

3.2 Relevant existing standards and initiatives

In order to make data interoperable, standards must be developed, adopted and implemented. The data submitted for the NEFIS project were too heterogeneous to enable them to be made interoperable (see section 3.6). Therefore development of the NEFIS metadata focussed on the metadata for cataloguing and identification of resources.

A review of the current status of standardization in the GI (Geographic Information) sector was carried out in the initial stages of the NEFIS project (Voss,

2003). The descriptions of the three geospatial metadata standards in this section (ISO 19100 series, FGDC, Open Geospatial Consortium) are based on this review.

Electronic searching and exchange of metadata require standardization. There are actually numerous approaches and standards for organizing data and information resources using metadata. All these initiatives follow common principles of providing a set of standards which allow for discovery of information resources across domains based on a common metadata framework tailoring to particular needs and data and information types.

3.2.1 Dublin Core

The Dublin Core Metadata Initiative (DCMI) is an organization dedicated to promoting the widespread adoption of interoperable metadata standards and developing specialized metadata vocabularies for describing resources that enable more intelligent information discovery systems. It was originally developed for author-generated description of web resources. However in the recent past it has received increased attention also from other communities such as museums, libraries, government agencies, and commercial organizations. The main DCMI aims are to simplify the identification of resources in the Internet through developing metadata standards for the discovery of information resources across subject domains and by defining frameworks for the interoperation of metadatasets. The Dublin Core Metadata Element Set (DCMES) consists of a stable set of 15 elements. DCMES has been published as ISO standard 15836 (ISO, 2003a).

3.2.2 ISO 19115: Geographic information – metadata

The International Organization for Standardization (ISO) is a worldwide federation of national standards bodies from more than 140 countries, one from each country. ISO is a non-governmental organization established in 1947. The mission of ISO is to promote the development of standardization and related activities in the world with a view to facilitating the international exchange of goods and services, and to developing cooperation in the spheres of intellectual, scientific, technological and economic activity. The products of the ISO work are international agreements which are published as International Standards.

During the course of the NEFIS project, ISO 19115 "Geographic Information – Metadata" has been published. ISO 19115 is part of the family of ISO Technical Committee TC211 (Geographic information/Geomatics) standards, and national profiles of the standard have been (or are being) developed and applied for particular countries, and adopted by initiatives such as INSPIRE. ISO 19115 defines over 300 metadata elements in 14 categories covering information about the identification, the extent, the quality, the spatial and temporal schema, spatial reference, and distribution of digital geographic data (ISO, 2003b). The encoding rules and XML schema standards for implementing the 19115 metadata will be covered in ISO standard 19139 "Geographic Information – Metadata – XML schema implementation". Future developments of a forest information system should ensure compliance with these standards.

3.2.3 INSPIRE

In 2004 the European Commission adopted a proposal for a directive of the European Parliament and of the Council for establishing an INfrastructure for SPatial InfoRmation in Europe, or INSPIRE (COM(2004) 516 final). The INSPIRE Directive was published in the Official Journal of the European Union on April 25, 2007 and entered into force on May 15, 2007 (EC, 2007). The Directive lays down general rules for the establishment of a European spatial data infrastructure, for the purposes of environmental policies and policies or activities which may have an impact on the environment (EC, 2004c). Many of the main principles have direct consequences for the development of metadata:

- Data should be stored, made available and maintained at the most appropriate level;
- It should be possible to combine spatial data from different sources across the Community in a consistent way and share them between several users and applications;
- It should be possible for spatial data collected at one level of public authority to be shared between all the different levels of public authorities;
- Spatial data should be made available under conditions that do not restrict their extensive use;
- It should be easy to discover available spatial data, to evaluate their fitness for purpose and to know the conditions applicable to their use.

For the documentation of data the following standards are to be used in order to comply with INSPIRE specifications (INSPIRE, 2002):

- ISO/TS 19103 Conceptual schema language;
- ISO 19109 Rules for application schema;
- ISO 19110 Feature cataloguing methodology;
- ISO 19115 Metadata;
- Dublin Core metadata standard for information discovery.

3.2.4 Open Geospatial Consortium

The Open Geospatial Consortium, Inc. (OGC) is a non-profit, international, voluntary consensus standards organization that is leading the development of standards for geospatial and location based services. Through its member-driven consensus programs, OGC works with government, private industry, and academia to create open and extensible software application programming interfaces for geographic information systems (GIS) and other mainstream technologies. With regards to metadata the OGC has adopted ISO 19115 replacing the OGC Abstract Specification Topics 9 (quality) and 11 (metadata). Additions to ISO 19115 were adopted in the course of further elaboration of metadata standards (www.opengeospatial.org/standards/).

3.2.5 Content Standard for Digital Geospatial Metadata

The Content Standard for Digital Geospatial Metadata (CSDGM) has been developed by the United States Federal Geographic Data Committee (FGDC). It was designed to "provide a common set of terminology and definitions for the documentation of digital geospatial data" (FGDC, 1998). CSDGM version 2.0 has been mapped to ISO 19115 (Intergraph_Corporation, 2003). In 2004 the American National Standards Institute adopted ISO 19115 as a national standard; FGDC is developing the national profile of ISO 19115 and CSDGM version 3.0 will be based on this profile.

3.3 Controlled vocabularies

The aim of producing a controlled vocabulary for the NEFIS project was to enable the data/resource to be catalogued by the owners and identified and retrieved by the searchers using a common list of consistently applied terms. The aim of using the controlled vocabulary is to improve both recall (i.e. the proportion of relevant material contained within a system that is actually returned by a particular search) and precision (i.e. the proportion of the retrieved material that is actually relevant to the searcher). In order for recall and precision to be improved, both the cataloguers and searchers have to be familiar with the terms used.

The current trend in terminology work is towards development of ontologies for the semantic web. However sound development of ontologies needs to be based on sound terminology, thesauri and classifications including definitions, relationships and structure.

- Ontologies are natural successors of thesauri, particularly for information retrieval and knowledge management.
- Ontologies provide better semantic representation and machine interpretable representation of knowledge. They are meant both for human as well as machine use.
- Transforming existing thesauri into ontologies brings increased precision of semantics particularly for information retrieval purposes and increased flexibility.

The parallel developments of internationally recognized thesauri illustrate this trend. They are for example:

- General Multilingual Environmental Thesaurus (GEMET) (www.eionet.europa.eu/gemet/): GEMET has been developed in all languages of the EU-member states within the working programme of the European Topic Centre on Catalogue of Data Sources (ETC/CDS) European Environment Agency (EEA) (Batschi et al., 2002).
- FAO Terminology (FAOTERM) (www.fao.org/faoterm/index.asp?lang=EN): FAOTERM is a multilingual (Arabic, Chinese, English, French, Spanish) terminology that aims to harmonize and standardize the terms used in

FAO documents and publications (agriculture, biology, forestry, fisheries, economics, statistics, nutrition).

- AGROVOC (www.fao.org/agrovoc/): AGROVOC is a multilingual (Arabic, Chinese, Czech, English, French, Portuguese and Spanish) agricultural thesaurus. AGROVOC is used for the description of information sources in the field of agriculture, fisheries, forestry, nutrition, food safety, and related subjects, such as the environment.
- FRA2005 Terms and Definitions (www.fao.org/forestry/site/13637/en/): A website containing terms and definitions used in the Global Forest Resources Assessment 2005 (FRA 2005). These definitions build on earlier global assessments to ensure backward comparability. They have been modified on the basis of recommendations from experts in various fora. The definitions have been translated (Arabic, English, French, Spanish and Russian) for the documents that guide the FRA national correspondents.
- SilvaVoc (www.iufro.org/science/special/silvavoc/): SilvaVoc is IUFRO's clearinghouse for a multilingual forest terminology. Through the harmonization of existing terminological data worldwide, the group aims to provide information on dictionaries, glossaries and terminological publications in forestry in different languages. The website also contains links to SilvaVoc's involvement in a bibliography of terminological publications, the SilvaTerm database, the multilingual glossary of forest genetic resources, a joint FAO/IPCC/CIFOR/IUFRO harmonization project, and other related projects. The site is available in English, French, Spanish and German.
- CAB Thesaurus. The CAB Thesaurus is a controlled vocabulary resource of over 48 000 terms for the applied life sciences (including forestry and forest products).
- Agricultural Ontology Server (AOS) www.fao.org/aims/aos.jsp: The goal of the AOS initiative is to discuss common semantic standards for enhanced web applications and improving access to agricultural information resources.
- National Agricultural Library (NAL) Agricultural Thesaurus: http://agclass.nal.usda.gov/agt/agt.shtml The NAL thesaurus is an online vocabulary look-up tool for agricultural and biological terms.
- (Semantic) Knowledge Organization Systems www.comp.glam.ac.uk/pages/research/hypermedia/Facet/index.htm: The problems of keyword searching are well known. Significant differences in results stem from trivial variations in search statements. These problems can be alleviated by Knowledge Organization Systems (KOS), such as classifications, gazetteers, lexical databases, ontologies, taxonomies and thesauri. KOS model the underlying semantic structure of a domain for the purposes of document retrieval. They act as semantic road maps and make possible a common orientation by indexers and future users –whether human or machine. A vast legacy of large multilingual vocabularies, indexed multimedia collections and indexed print collections is available. They exist in a network of practice, education, training and mechanisms for evolution (Tudhope, 2004).

A related topic is the use of classification and coding schemes. Such schemes are well known in the library and information community and are becoming more popular for cataloguing and retrieving electronic information. A Global Forest Decimal Classification (GFDC) was published by IUFRO (Holder et al., 2006). The GFDC replaces the former Forest Decimal Classification (FDC) and Oxford System for Decimal Classification for Forestry (ODC), and is the official expansion of the Universal Decimal Classification (UDC) for forestry. Currently the classification is bilingual (German and English) and there are plans to publish French and Spanish editions.

A recent development in the use of metadata to catalogue information is the development of folksonomies. The term 'folksonomy' was originally used by Thomas Vander Wal. He defines it as:

> *...the result of personal free tagging of information for one's own retrieval. The act of tagging is done by the person consuming the information. The value in this external tagging is derived from people using their own vocabulary and adding explicit meaning, which may come from inferred understanding of the information/object as well as. The people are not so much categorizing as providing a means to connect items and to provide their meaning in their own understanding (VanderWal, 2005).*

Perhaps the best known example of this kind of collaboration is del.icio.us (a service for sharing and storing web bookmarks; http://del.icio.us). However, this has inspired some more academically oriented examples such as Connotea (www.connotea.org/), which is an online reference management and bookmarking service for scientists created by Nature Publishing Group (Lund et al., 2005), and CiteULike (www.citeulike.org/) which is a bookmarking service for academics and specializes in providing tools for organizing, storing and sharing citations.

3.4 NEFIS metadata schema development

3.4.1 The DC metadata schema elements

The work dealing with metadata within the NEFIS project consisted of two tasks. The first task was the elaboration of a metadata description format (or 'metadata schema') for information resources for application to the data available for the NEFIS project. The second task was to investigate the possibilities of establishing a set of controlled vocabularies (or subject key words) representative of the NEFIS data providers' datasets. The metadata schema development was to be based on existing standards as laid out in section 3.2. Building on gained experience in previous projects, it was proposed to develop further the generic DCMI standard to meet the particular data provider needs (JRC, 2002b). In order to better follow the process of the NEFIS metadata schema development, the DCMI and the Dublin Core Metadata Element Set (DCMES) will be introduced, thus providing the background for choosing DCMES as the basis for the NEFIS schema.

The DCMI is as mentioned a well established and recognized organization dedicated to promoting the widespread adoption of interoperable metadata standards while having interaction with other metadata initiatives. The DCMES comprises 15 elements, which have been developed through consensus by an international group of professionals from librarianship, computer science, text encoding, the museum community, and other related fields of scholarship.

As Table 3 shows the DCMES is a minimalist approach to metadata and identifies a stable core set of metadata elements to describe electronic information resources. They are not comprehensive nor subject-specific, but generic and can be used to provide a first level of description for almost anything in any subject domain. By this approach it stays rather simple and easy to grasp. This simplicity means that the metadata are relatively easy (and cheap) to produce, and that the metadata produced are commonly understandable across subject domains. However, it also means that the metadata does not support the semantic interoperability afforded by more complex metadata schemas. The DCMES is intended to be used in combination with other metadata standards.

Although the DCMES is not very comprehensive it is open to qualification, i.e. the addition of additional subject specific sub-elements. These could be for example domain specific qualifiers for agricultural, plant genetic or forestry information resources. In this way the 15 elements function as a core set that can be surrounded by different subject domain qualifications. This approach is embodied in what the DCMI calls 'the dumbing down principle'. It states that by ignoring subject domain qualifications and only searching on DCMES elements queries can be made across subject domains (DCMI, 2005a). Figure 17 shows the DCMES as being a common core through diverse subject domain qualification sets.

Table 3. The 15 core elements of the DCMES (DCMI, 2004; DCMI, 2005b).

Content	Intellectual property	Instantiation
Coverage	Contributor	Date
Description	Creator	Format
Type	Publisher	Identifier
Relation	Rights	Language
Source		
Subject		
Title		

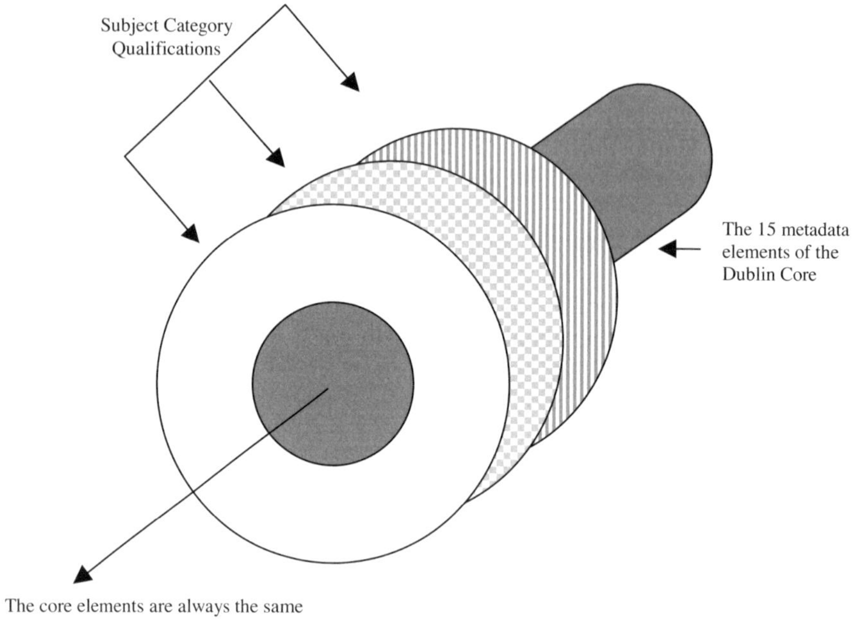

Figure 17. The Dublin Core metadata element set is a common denominator through multiple subject domains.

Further there is the option to use 'other elements', 'element refinements' or 'encoding schemes' in addition to the 15 core elements in the DCMES (DCMI, 2005b).

Additional elements (such as audience, accrual method, accrual periodicity, provenance) have been defined by DCMI in addition to the 15 core elements.

Element refinements can make the meaning of a DCMES element narrower or more specific. Such element refinements are a property of a resource or resources which shares the meaning of a particular DCMES element but with narrower semantics. For example, the DCMES element 'coverage' which describes the extent or scope of the content of the resource may need to include additional information on the exact spatial location (a place name or geographic coordinates), the temporal period (a period label, date, or date range) or jurisdiction(s) (such as a named administrative entity). The DCMES element 'rights' can be used to record information about rights held in and over the resource. For example, this could typically contain a rights management statement for the resource, or it could be a reference to a service providing such information. Rights information often encompasses intellectual property rights, copyright, and various property rights. This element may in addition have as refinements access rights, rights holder or licensing.

Encoding schemes provide contextual information or parsing rules that aid in the interpretation of a term value. Examples for encoding schemes include: (1) the

W3CDTF which are encoding rules for dates and times based on a profile of ISO8601 (international standard date and time notation); and (2) ISO3166 (country codes for the representation of names of countries).

A mapping between DC metadata elements (the 15 DCMES elements plus audience) and ISO 19115 has been carried out (CEN, 2003a; CEN, 2003b). There are four elements that have no correspondence with elements in the core version of ISO 19115: contributor; relation; rights; audience. However, these elements do correspond with one or more elements in the full version of ISO 19115. There are also elements in the core version of ISO 19115 that have no direct correspondence with elements in the DCMES.

3.4.2 NEFIS additions

Having provided a background to the DCMES and its 15 core metadata elements, we will now introduce the NEFIS metadata schema (Figure 18). The development of the metadata schema was implemented as a consultative process between NEFIS participants, system developers, information/data providers and users to include both additional main elements and a number of refinements (Schuck and Green, 2005).

In particular the issues of data quality, data types and access rights were given special attention. The actual vocabulary development under the DC element 'subject' was restricted to such topics which represented the data providers' datasets and expertise. Further additional metadata elements that were seen as useful for NEFIS were discussed within the project consortium, but not further developed. They included, for example, 'security', 'forest management models', and 'analytical tools'.

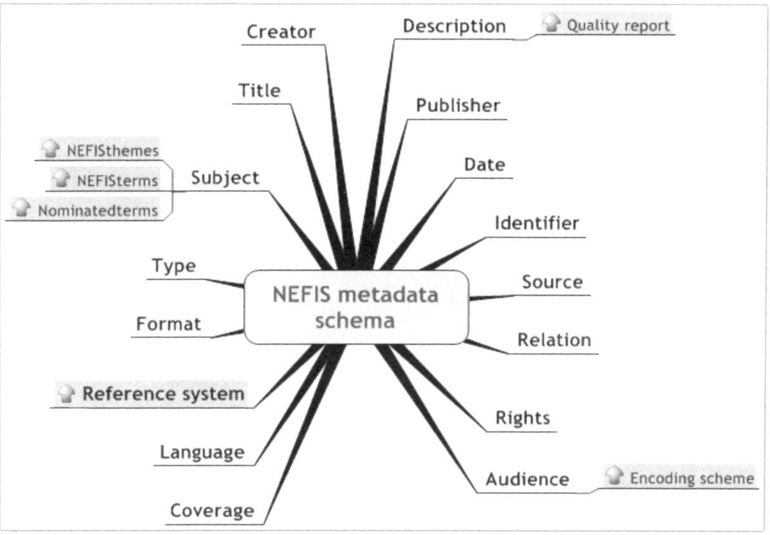

Figure 18. NEFIS metadata schema based on DCMES. NEFIS modifications (⇧).

3.4.2.1 Audience

'Audience' as a main metadata element was included as part of the NEFIS metadata schema. It is defined as a class of entities for whom the resource is intended or useful. The element was regarded as useful in order for data providers to give insight on their clients and/or key target groups. As the DCMI does not specify an encoding scheme for 'audience' the recommendation is to develop a controlled vocabulary for the specific use of this element (Table 4). In the NEFIS project this was based on a list developed under the EFIS project (JRC, 2002b).

3.4.2.2 Reference System

As the DCMES does not adequately cover geo-referenced datasets, and many of the project participants were willing to provide such datasets, the element 'reference system' was added to the NEFIS metdata schema. It was first suggested as an additional refinement under the element 'format' but due to its importance for geospatial datasets it was then given the status of a separate element in the schema. The approach used for the element reference system was based on the EEA metadata form for spatial datasets (EEA, 2003) which is derived from the ISO19115 (ISO, 2003b). For the NEFIS element 'reference system' the crucial descriptive components or decriptors were selected (e.g. EllipsoidName, ProjectionFalseNorthing) and are presented in Table 5.

Table 4. NEFIS encoding scheme for the element 'audience' (adapted from IMRC, 2004; JRC, 2002b).

Descriptor	
business	government
consultants	government agencies
donor organizations	media
educators	non-governmental organizations
environmental groups	policy-makers
financial institutions	public
forest landowners	researchers
forest managers	students
forestry associations	

Table 5. Descriptors under the element reference system (adapted from: EEA, 2003).

Reference System
ReferenceSystemName
DatumName
EllipsoidName
EllipsoidSemi-majoraxis
EllipsoidAxisUnits
EllipsoidFlatteningRatio
ProjectionName
ProjectionZone
ProjectionStandardParallel
ProjectionLongitudeofCentralMeridian
ProjectionLatitudeofProjectionOrigin
ProjectionFalseEasting
ProjectionFalseNorthing
ProjectionLongitudeofProjectionCentre
ProjectionLatitudeofProjectionCentre

3.4.2.3 NEFIS themes, NEFIS terms, Nominated terms

Under the DC element 'subject' there were three element refinements added: (1) the 'NEFIS themes' element refinement can be described as a set of broad topics within forestry and forest-related information; (2) the subject keyword lists or 'NEFIS terms' were developed either at a detailed or broad level; and (3) 'Nominated terms' was added to accommodate the possibility to submit terms and subject keywords not found in the 'NEFIS terms' controlled vocabulary.

From the beginning of the project it was clear that the development of a full controlled vocabulary for European forestry was well above the resources available within the project. An attempt was made to explore different approaches and possible solutions for compiling a controlled vocabulary. The work followed closely from existing, recognized vocabularies (see section 3.3) and took into consideration vocabularies used by the NEFIS partners where these were available.

The processes of establishing the NEFIS themes and the corresponding controlled vocabularies or subject keyword lists under these themes were investigated using a cooperative consultation process between data providers/users and ontology and library experts. In a first step, NEFIS partners identified 19 'themes' (Table 6). The 'NEFIS themes' are not to be seen as a new classification within the forestry domain, but as a contribution to ongoing activities in vocabulary and thesauri development. Definitions for the themes and the number of terms in the list are also shown in Table 6.

Table 6. NEFIS Themes, definitions for the themes, and number of main terms in the list.

NEFIS Theme	**Definition**	**Number of main terms**
Forest inventory	This theme includes growth and yield (including mensuration); forest resource inventory (collection and analysis of resource data); forest management planning and managerial economics; remote sensing; management sciences of forest enterprises; statistical methods, mathematics and computer technology	212
Forest health	This theme includes physiological and genetic interactions between trees and harmful biotic impacts, including resistance mechanisms: biological and applied aspects of tree diseases; environment/pathogen interactions in forest decline; the biology and control of forest tree insects; impacts of air pollution on forest trees and forest ecosystems, including diagnosis, monitoring, biology, genetics and treatment of polluted forests	280
Silviculture	This theme includes forest and ecosystem management; stand establishment and treatment (including forest nurseries); agroforestry; biomass for energy; restoration of degraded sites; mountain zone and arid zone silviculture; tropical, boreal and temperate zone silviculture; and natural (extensive) and artificial (intensive) silvicultural systems	458
Vegetation	This theme includes descriptions of forest types, vegetation types and soil types.	512
Non-wood goods and benefits	This theme includes goods of biological origin other than wood, derived from forests, other wooded land and trees outside forests and non-wood benefits, including cultural, social and spiritual benefits, and employment and community benefits.	27
Forest operations	This theme comprises forest engineering (including building, construction, and machinery and operational methods in all forestry practices from stand establishment to harvesting and wood delivery); operational planning and control.	87
Forest fire	This theme includes forest fire prevention and control; and the use of fire as cultural tool.	47
Forest products and trade flows	This theme includes the fundamental nature of wood and products derived from wood, their protection in storage, production, utilization and trade.	358

Table 6. Continued.

Rural development	This theme includes all aspects of rural development in the context of forestry and forest policies, including forest recreation and landscape management; the social and institutional aspects of forestry.	63
Forestry institutions	This theme includes educational, governmental non-governmental and commercial organizations working in forestry and related areas.	23
Maps and geo-referenced data	This theme includes geographic information systems (GIS) data structures for both raster and vector digital maps, georeference and terrain data, cartography, maps, and image analysis.	119
Field Experiments	This theme includes experiments on forest ecosystems, forest stands, trees, tree components or other parts of the forest ecosystem. Forest field experiments are conducted on-site to study the biological response to different, controlled treatments (in contrast to, for example, laboratory experiments and forest engineering and work science studies). Generally, field experiments are conducted to test or examine specific scientific hypotheses. The duration of field experiments varies depending on forest type and research objectives. So-called long-term field experiments often comprise the entire rotation or a considerable part of the life-span of trees.	281

Partners were then asked to submit keyword terms for the themes for which they could provide expertise. No keywords were submitted for some of the themes (environment and ecology; tree physiology and genetics; forest economics; forest labour; forest policy; forest research; wood fuel).

The resulting keyword lists were variable in the level of detail and scope. This result reflects the 'real world' and the fact that resources were not available for a time-consuming collaboration for development of a high quality product. In parallel with this effort, terminology experts identified existing relevant keyword lists that could be used as a base for the NEFIS vocabulary and developed detailed keyword lists (forest inventory; silviculture; forest products and trade flows). When the partner's datasets were compared with their individual use of keywords, the added value of continuous feedback by the partners became very clear: better integration of individual datasets, and therefore better search and retrieval outcomes would be possible in the long run.

To overcome inconsistencies, attention was given to evaluating the requirements of granularity of a NEFIS keyword list. In other words, to determine what level of detail the keyword list would need in order to satisfy the needs of the NEFIS searchers and NEFIS cataloguers. Multilingualism and improved semantics through the incorporation of definitions for the keywords presents three distinct advantages:

Table 7. Excerpt from the keyword list for the NEFIS Theme Forest inventory. In addition to these categories, some terms contain additional information concerning plural forms, short forms, and synonyms. CABI_BT= CAB Thesaurus Broader term. FDC=Forest Decimal Classification.

English	CABI_BT	Categories-FDC	Categories -IUFRO	En_Definition	En-Note
deforestation	environmental degradation		3.2.3. Harvesting/ Utilisation	The long-term removal of trees from a forested site to permit other site uses.	Cutting of trees followed by regeneration is not deforestation.
forest area		615	2.1.3. The Organisation of Forests	The total area of a forest enterprise or management unit, i.e. the sum of the productive and non-productive forest land.	
rotation age	TBFRA	613	1.3. Spatial and temporal organisation of forests	The planned number of years between the establishment or regeneration of a tree crop or stand and its final cutting at a specified stage of maturity.	The age at harvesting when it coincides with the rotation. If the age does not coincide with the rotation, it is termed felling age. Stand and forest average rotation ages may differ. Also, there is no real rotation age for uneven-aged stands and forests. (Source: Forest Management Terminology)

English	French	German	Hungarian	Italian	Spanish
deforestation	déboisement	Entwaldung			
forest area	surface boisée	Forstbetriebsfläche	erdögazdasági terület	superficie forestale totale	superficie de un predio forestal
rotation age	age de révolution	Umtriebsalter	vágásérettségi kor	età al taglio	edad del turno

Table 8. Excerpt from the NEFIS Theme Forest health.

NEFIS Theme: Forest health	
abiotic damage	plant condition
anthropogenic damage	forest dieback
biological control	forest decline
biotechnical control	hail damage
biotic damage	ice damage
controlled fire	integrated control
disease control	lightning damage
forest damage	mechanical control
forest fire	nursery diseases
forest health	plant pathology

(1) improved comprehension of contextual terminology; (2) better thesaurus structures; and (3) overall, as a demonstration that the role of the service is also to educate through improved communication and demonstration.

Table 7 shows an excerpt of the keyword lists for a theme developed to a detailed level (forest inventory). Table 8 shows an excerpt from the 'flat' list of keywords produced for the theme 'forest health'.

The resource discovery component of the advanced NEFIS demonstrator was populated with the list of English terms only. Further development of the resource discovery would be needed to implement the advanced features of the keyword lists (i.e. the definitions, notes, relationships with other terminologies/classifications, equivalent terms in other languages).

In the process of development of the subject keyword lists, the project partners felt that there should be an avenue to submit terms not found in the NEFIS controlled vocabulary, but are of crucial importance to describe a particular resource. In vocabulary development it is always a delicate balance to produce a comprehensive set of controlled keywords for different subject themes without them becoming too extensive and unwieldy. Very exhaustive lists are difficult to apply by data providers, and may become overwhelming for information searchers. The option to submit 'Nominated terms' can allow current lists to be updated and were seen as highly useful by the project partners during the development of the controlled vocabularies. The disadvantage of allowing use of terms not in the controlled vocabulary is that the keyword vocabulary could become littered with non-standard terms (mis-spellings, synonyms or near synonyms for which there is a term in the controlled vocabulary) and interoperability of the system would be affected as non-standard terms are added.

An information searcher using the advanced NEFIS demonstrator has two practical options to specify a metadata search with regards to key terms (Figure 19). It can either be done as a 'free text search', meaning a searcher may enter their

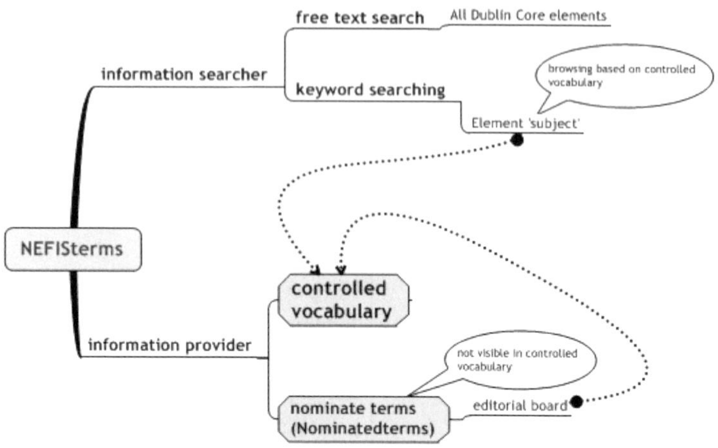

Figure 19. Options for cataloguing a resource using a controlled vocabulary (NEFIS terms) and nominated terms, and searching the catalogue using the controlled keywords or free text.

own terms into the search interface. Free text search will then search all the metadata elements and their refinements. The 'controlled term search' function allows the user to select from a controlled set of subject keywords. The search is then made of the NEFIS themes and NEFIS terms. Nominated terms however are not visible to a user searching the NEFIS terms list.

It was suggested that the Nominated terms refinement should only be for the data provider who may use this option when describing an information resource and should not be searchable by users of the system. This would allow for the possibility to screen the terms and eliminate non-standard terms. Nominated terms would in a first step be subject to review by an editorial board, and if found suitable would be added to the NEFIS terms controlled vocabulary list under the respective NEFIS theme. If there are synonyms, or errors (e.g. spelling mistakes) then the suggested term could be cross referenced to the preferred term or deleted.

One possibility for an operational forest information system could be an automatic system for monitoring the frequency of entries of 'new' terms added by data providers to the Nominated terms field, and thus assist an editorial board in its work. The editorial board would review the frequency with which new search terms, for example 'genetically modified trees', appear and then decide if the term should be moved to the controlled vocabulary. Conversely, terms in the controlled keyword lists that are rarely or never used by data providers could be removed and stored in a 'keyword repository'. Tracking the origin of the nominated terms is a key issue that was discussed within the NEFIS project, but not implemented in the advanced version of the metadata catalogue and resource discovery component. In addition to the new term itself, those suggesting new terms could be prompted for additional information, such as definitions, equivalences in other languages, with references to sources of this information if it is available.

An issue that has to be addressed is the interoperability of the developed vocabularies with other forest information systems. The Global Forest Information Service (GFIS) has as one of its key aims and needs the development of a multilingual forestry thesaurus that will be designed to enhance the effectiveness of GFIS search operations and guarantee improved access to primary forest information and other forestry information systems. It is envisaged that it will be developed with close attention to existing internationally recognized thesauri and ontologies to ensure coherence with internationally accepted standards. The NEFIS project has progressed in close association with GFIS, sharing many metadata features and standards, and the keyword lists developed within the project will be useful inputs in any future GFIS thesaurus work. GFIS is actively engaged in developing and defining metadata standards for many different resource types, and NEFIS datasets could be viewed as a subset of the types that will eventually be retrieved through GFIS, with similar technical and semantic characteristics. NEFIS keyword lists have very little structure at present for reasons discussed above; they will therefore be readily usable by any number of different systems.

3.4.2.4 Quality Report

The most debated topic of the NEFIS metadata schema was that of quality. It was agreed that a general indication of quality is inherent to the DCMES, their element refinements, and encoding schemes. The project participants expressed however the need for specifying in more detail the quality of a submitted information resource, and therefore provide the users with a clear assessment of quality. The approach for adressing this need was the establishment of a 'quality report' element refinement under the DC element 'description'.

Providing an objective assessment of the quality within the metadata is complex but important. There is the issue of the quality for which data were created in the first place, but also if the aim is to allow reuse of the data, the question arises of how well suited the data are for other uses.

An adequate description of quality of a resource is of the utmost importance. This is relevant both for the data provider who may want to as complete a description of a resource as possible, and for the informaton searcher who will want the best possible indication of quality before investing further time (and money) into a potentially interesting resource. When considering the DCMES and its refinements and encoding schemes, there are a number that provide an indication of quality. The metadata elements of main importance are: creator, description, publisher, coverage, source, date and audience (Box 1).

Box 1. DCMES elements indicating quality of a resource (DCMI, 2005a).

- *Creator*
  The creator is an entity primarily responsible for making the content of the resource. Examples of a creator can include a person, an organization, or a service
- *Description*
  The description is an account of the content of the resource. It may include but is not limited to: an abstract, table of contents, reference to a graphical representation of content or a free-text account of the content.
- *Publisher*
  The publisher is an entity responsible for making the resource available. Examples can include a person, an organization, or a service.
- *Coverage*
  The coverage is the extent or scope of the content of the resource. It will typically include spatial location (a place name or geographic coordinates), temporal period (a period label, date, or date range) or jurisdiction (such as a named administrative entity).
- *Source*
  The source can be a reference to a resource from which the present resource is derived. This means that resource which is actually being described may be derived from the 'source resource' in whole or part.
- *Date*
  The date gives information associated with an event in the life cycle of the resource. Normally the date will be associated with the creation or availability of the resource. It can also infdicate the last event of updating the resource which is relevant for websites and databases.
- *Audience*
  The audience can give an indication towards the class of entity for whom the resource is intended or useful. Such a class of entity may be determined by the creator or the publisher or by a third party. Examples may be target groups of the resource (educators, researchers, non-governmental organizations, forestry associations)

This was reflected by the NEFIS evaluation of the proposed metadata schema where most consortium partners considered that quality aspects were addressed –at least to some extent –within the DCMES (Figure 20).

The issue of quality is addressed in the ISO 19115 standard. One of the top level categories of the metadata is data quality. The elements addressed within the category include attribute accuracy, logical consistency, completeness, positional accuracy, lineage (source information and information about the processes/ transformations the data went through). Quality of geographical information is also addressed by other ISO standards: ISO 19138 Geographic Information – Data quality measures; and ISO 19113 Geographic Information – Quality principles; and ISO 19114 Geographic information – Quality evaluation procedures. Any further development of the metadata schema as implemented in NEFIS should take ISO

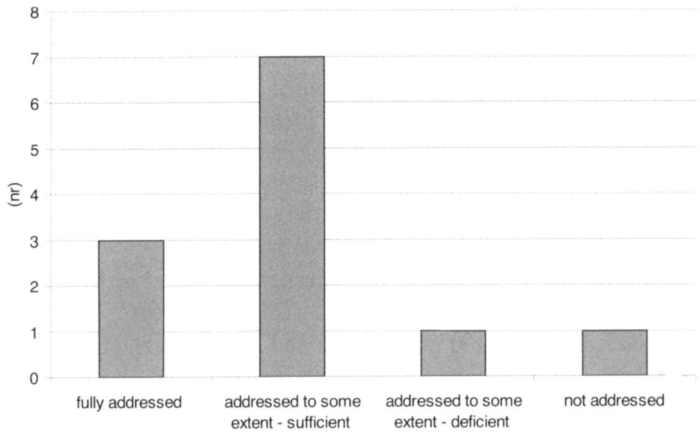

Figure 20. Incorporated quality aspects within the Dublin Core metadata element set.

standards as mentioned above into account and ensure conformity with these standards.

The NEFIS partners, however, expressed a desire to be able to specify the quality of a submitted information resource in more detail, and therefore provide users with a comprehensive, transparent, and objective assessment of quality. Therefore, the partners proposed that a quality field should be included as a DC refinement and that the usefulness of the refinement should be assessed.

Quality can be expressed in terms of statistical validation, scale and geographic accuracy, resolution, or by giving details on sampling and inventory methods, of how the data are organized, the existence of control protocols, and how often the data are updated. Following a lively debate on quailty, the following main headings for a 'quality report' were proposed: (a) collection mandate; (b) availability of data collection and data processing guidelines; (c) definitions; (d) sampling methods; and (e) explanatory notes. It was agreed that the information on quailty under the headings a-e would best be added under the DC element 'description' as an element refinement entitled 'quality report'.

The heading 'collection mandate' adds a political dimension to the data/information presented by a data provider. It will give a general indication on the authority in which an organization or individual is acting and whom they represent. For example, the UNECE/FAO Timber Committee provides forestry information for the Economic Commission for Europe region and acts within the framework of the policies of the United Nations based on agreements with country governments. It is stated in the terms of reference of the United Nations Economic Commission for Europe that it may "undertake or sponsor the collection, evaluation and dissemination of such economic, technological and statistical information as the Commission deems appropriate" (UNECE, 2005). The Finnish Forest Research Institute is commissioned by the Finnish Ministry of Agriculture and Forestry to implement the national forest inventory for Finland and make available its results. The availability of 'data collection and data processing guidelines' can be highly supportive to the interested user for understanding of data or datasets. They will give insight to the activities of the data provider concerning a particular resource. The approach of presentation can either be a textual description of the actual guidelines or the provison of a link to the full documentaion or the authority where they can be retrieved. Many national and interantional data collection agencies have such information at their disposal. A common difficulty here however will be that of language. The availability of guidelines is usually closely linked to that of 'definitions'.

There are numerous initiatives at the international level to harmonize or use commonly accepted definitions (COSTE43, 2004; EC, 1997; FAO, 2002; FAO, 2004; FAO, 2005; MCPFE, 2003a; Päivinen and Köhl, 2005). However there may still be differences due to the nature of the reseach or policy needs although some consesus with regards to thresholds may have been reached or is ongoing (Table 9). At the national level, definitions for forest area can still vary considerably from country to country (Table 10).

Table 9. International forest area definitions (parameters and their values); adapted from (FAO, 2002).

	minimum width of tree covered area [m]	minimum crown cover [%]	minimum area [ha]	minimum tree height [m]
FAO/UNECE, 2000	20	10	0.5	5
Worldbank Group	-	10	1	2
UNFCCC, Kyoto	-	10–30	0.05–1	2–5
UNEP/CBD/ SBSTTA, 2001	-	10	0.5	5

Table 10. National Forest Area Definitions (EC, 1997; Päivinen and Köhl, 2005).

Country	min width [m]	min crown cover [%]	min area [ha]	min production [m³/ha/year]
Denmark	20	30	0.5	
Germany	10	–	0.1	–
Finland	–	–	0.25	1
France	15	10	0.05	–
Greece	30	10	0.5	–
Ireland	40	20	0.5	4 (coniferous) 2 (broadleaf)
Italy	20	20	0.2	–
Netherlands	30	20	0.5	–
Austria	10	30	0.05	–
Portugal	15	10	0.2	–
Sweden	–	–	0.25	1
Switzerland	25–50	20	–	–
Spain	20	5	0.2	–
UK	50	20	2	

For such reasons a statement and if possible the listing of or link to definitions should be a central part of the quality report. A short introduction of the 'sampling method(s)' will allow insight into the methodological approaches that are applied to generate base data. For example, this could apply to national forest inventories plot design, the design of forest condition monitoring networks, or biodiversity related data collection procedures. Finally 'explanatory notes' can serve as a pool for describing resource specific details which have not appeared under the above headings. It could, for example, be used to explain how datasets based on national definitions have been adjusted to meet an internationally agreed definition or which institutions have been involved in the data collection, analysis and publication process. Examples of quality reports are presented in Box 2.

Box 2. Examples of quality reports

Title: "EFI-WFSE Forest products trade flow database"
NEFIS Theme: "Forest products and trade flows"
Source: "COMTRADE database of the United Nations Statistics Division (UNSD)"
Quality report:
Explanatory notes: The source data was purchased from United Nations Statistics Division (UNSD).
Availability guidelines: A cleaning process is applied to the source data to: (a) represent FAO classification of forest products; (b) standardize quantity units; (c) replace missing quantity data and (d) estimate missing trade flows. A full description of the data processing (standardizing quantity units, estimating missing quantity data, estimating missing trade flows, changing national and regional boundaries, and the database structure) can be found at http://www.efi.int/efidas/fpstf.html (references section). EFI/WFSE Trade Flow Database General description. Bruce Michie and Philip Wardle. European Forest Institute, Joensuu, Finland, 2000. Updated 2002.16p
Definitions: FAO Classification and Definitions of Forest Products (SITC1, SITC2, SITC3 http://unstats.un.org/unsd/cr/registry/default.asp, HS88, HS96 http://europa.eu.int/comm/eurostat/."

Title: "Long-term European forest resources assessment database"
NEFIS Theme: "Forest inventory"
Source: "Various forest resources assessements implemented by the FAO and UNECE/FAO (since the 1950s)"
Quality report:
Collection mandate: "Utilisation of international forest resources information officially published by FAO and UNECE/FAO. The information collected by FAO and UNECE/FAO is based on data questionnaire returns from designated national country correspondents. The data presented in the FAO and UNECE/FAO publications was transferred to electronic format and organised in an interactive Internet database."
Definitions: The definitions of the variables, reference years when data was assessed and units are included in the database structure and can be accessed at any time after conducting a query. The definitions applied are those used in the individual FAO and UNECEFAO publications. Note that definitions have been subject to change between reports.

Box 2. Continued.

Explanatory notes: "Take note that country names and borders may have changed over the period covered by the LTFRA database. The countries listed in the coverage>spatial refinement are those available from the latest forest resources assessment (UN-ECE/FAO, 2000. Forest Resources of Europe, CIS, North America, Australia, Japan and New Zealand: Main Report. Geneva Timber and Forest Study Papers, No. 17. 445 p. The LTFRA database also gives reference to the original publications. In the case of the latest forest resources assessment (2000) a link is given to the on-line report at the website of the UNECE Timber Committee."

Title: "Forest map of Europe"
NEFIS Theme: "Forest inventory"
Source: ""
Quality report:
Collection mandate: ""
Guidelines:
Definitions: "Various sources of data were used to compile the map according to the availability of data: (1) national forest inventory statistics at the sub-national level and national definitions were used for most EU 25 countries (if not available countrywide inventory or UNECE/FAO data were used), the European part of Russia, Norway, Switzerland, Bosnia and Herzegovina, Bulgaria, and The FYR of Macedonia. National definitions of forest area and species groupings apply; (2) Data and definitions from the Temperate and Boreal Forest Resources Assessment 2000 (implemented by the UN/ECE/FAO) were used at the country level for Albania, Belarus, Croatia, Iceland, Liechtenstein, Moldova, Romania, Slovakia, Slovenia, Ukraine and Serbia and Montenegro."
Sampling methods: "Step 1: the percentage of the forest proportion was estimated for each AVHRR pixel (1x1km), using CORINE land-use classification as training data to establish the link between five classes (forest, other wooded land, and within the forest class, coniferous, deciduous, and mixed forest classes) and the AVHRR spectral response. Step 2: the area of classes was calibrated to correspond to the utilised forest statistics within the given polygons (national or sub-national). . In the case of EU15 the NUTS administrative boundaries (Nomenclature of territorial unit for statistics) were used. For other countries national administrative boundaries (either sub-national or national) were used. A timberline mask was implemented to exclude areas considered above the timberline from the calibration process. Such areas were automatically assigned a 'zero' value."

In addition to the qualitative indicators, a set of quantitative measures of quality for some main target variables was proposed for forest inventory data. This included indicators such as the availability of standard error, the total sample size, and resampling for measurement quality control (i.e. the percentage of plots measured by a checking crew; see Box 3 for an example).

3.4.2.5 Access rights

A central theme in NEFIS has been to address the topics of data rights and access based on the precondition that the actual data will reside with the providers. The concerns of the actual data providers (administrations, organizations, networks, individuals) towards access rights were quite diverse relating to their individual data

Box 3. Quantitative quality measures for forest inventory data

Availability of standard error: Yes
Variable name: Total volume
Standard error: ca. 1.5–2.5% (varies between regions)
Total sample size: 65859
Sampling unit: plot
Resampling: Yes
Resampling percentage: ca. 1%

policies. Many organizations, may have in place policies on access rights, which in cases may be linked to actual copyright issues.

When looking at the needs of an operational system there will be several layers of access as related to the type of user. One could call that role based access rights. It may mean that a certain user group may be allowed only to access a subset of the entire dataset/database, another user community may need to register for full access, and a third group may have full access rights for both using and adding/editing data. Such role based access rights will need to be taken into account, both by the data provider and the system designers, when setting up an overall forest information system. The same conditions can apply if organizations ask for datasets to be hosted and maintained on a central server. A designated body will have maintenance rights and will guarantee the running and availability of the hosted data (no rights to editing the data contents). That body will also ensure that access rights policies as requested by data providers are respected.

This will mean that for an operational information system there is a need for tailored access rights combined with a potential mix of centralized and distributed data hosting. Therefore, there will not be one standard solution to the issue of data access control. When developing the overall system there will be either the need to make available a set of options to data providers (access rights packages) or give clear guidance on how to set up tailored access restrictions.

For example, an operational forest information system would allow access to a centralized database where the actual dataset is maintained. In agreement with the data holders the database host is commissioned to grant access options to user communities according to their particular user profile, including privileged or general users (Figure 21).

This means that access can be granted either directly to all or parts of that database, or support particlar user groups with tailored services (reporting tools, Web services). Users communities with different access permissions that can be envisaged may be: the European Commission and its services (EC-DG's, EUROSTAT); mandated data collection agencies (Regional FRA activities); the scientific community (research organizations, research projects, individual researchers and stakeholders); or the general public.

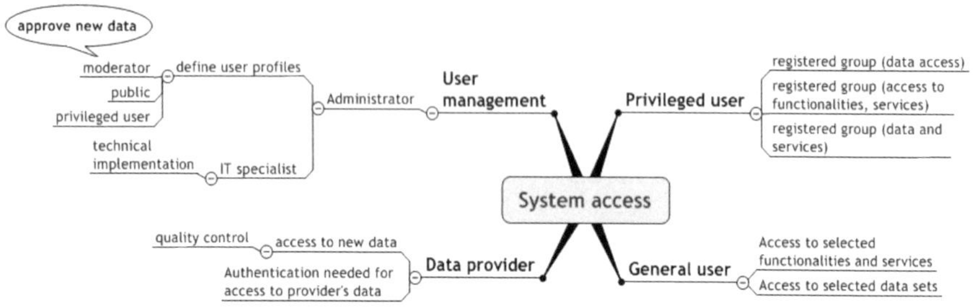

Figure 21. System access: different actors within an information system.

3.5 Evaluation of the proposed NEFIS metadata schema

The NEFIS project has concentrated a major part of its work on the improvement and development of the NEFIS metadata schema for discovery of information resources. In particular, finding a balance between the content, structure and usability of the metadata schema is necessary to ensure that the various data providers will find added value for metadata and data provision through an operational European forest information system.

An evaluation of the project deliverables was coordinated by the NEFIS partner from the University of Hamburg. With respect to the metadata schema, the evaluation involved a questionnaire survey concerning: (a) the usability and applicability of the metadata schema for describing datasets; and (b) the data rights and data rights management. The questionnaire was completed by the NEFIS partners.

Although more than 75% of the questionnaire respondents already had contact with metadata in some form the knowledge and experience using and dealing with metadata can only be described as average. Most partners (~70%) considered the workload and complexity to prepare and enter metadata records for their datasets as acceptable (Figure 22). On average partners spent more than five hours on the process of understanding the metadata schema, preparing and searching, and then compiling and recording the required metadata.

Simplicity in recording the required metadata information is fundamental for making the schema successful and accepted by various user communities. Therefore it was of interest within the evaluation to identify those elements where partners had problems with respect to applicability and usability. Applicability and usability can be described by various aspects – for example, workload per element, understanding of element requirements, but also of the actual relevance of proposed metadata elements. Table 11 provides a summary of the results of the NEFIS evaluation. It highlights which of the proposed metadata elements were, for most partners, challenging/problematic or easy/acceptable to apply. Three categories of applicability and usability are distinguished:

a) *easy:* elements with which partners had no problems
b) *acceptable:* elements where some partners had minor problems
c) *difficult:* elements with which many partners had severe problems ('hot spots')

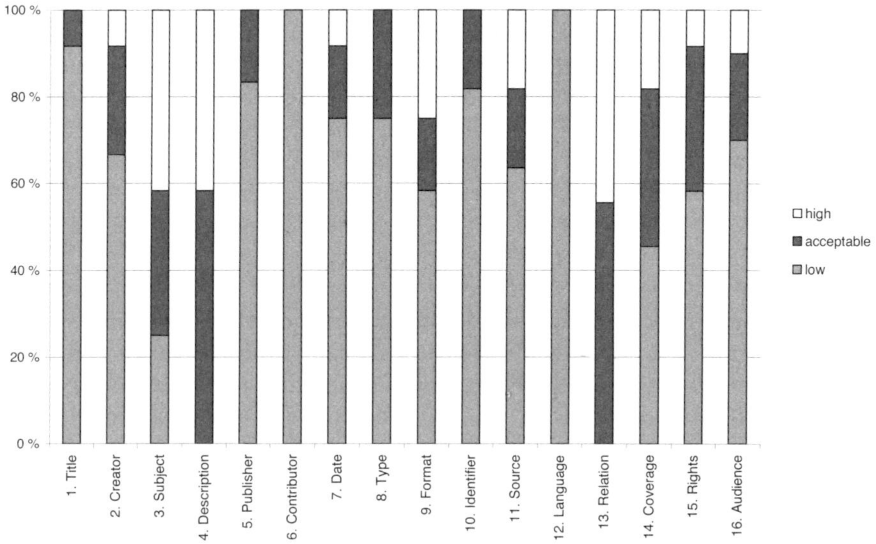

Figure 22. Workload – time to prepare and enter required information.

Table 11. NEFIS metadata schema – summary of applicability and usability. ☺=easy; ?=acceptable; ??=difficult (hotspots).

Metadata Element		Evaluation
1.	Title	☺
2.	Creator	?
3.	Subject	??
4.	Description	??
5.	Publisher	☺
6.	Contributor	?
7.	Date	☺
8.	Type	☺
9.	Format	??
10.	Identifier	?
11.	Source	?
12.	Language	☺
13.	Relation	?
14.	Coverage	☺
15.	Rights	?
16.	Audience	?

The elements 'subject', 'description' and 'format' caused most problems when preparing metadata records. Especially the element 'subject' with its refinements 'NEFIS themes' and 'NEFIS terms', but also compilation of the element refinement quality report (under the DCMES 'description'), were seen as challenging to complete. Possible reasons for the difficulties could be: (a) inexperience with metadata preparation; (b) the complex structure of the metadata schema itself; or (c) ambiguous descriptions in the metadata guidelines. However, both subject and description were regarded as crucial pieces of information within the metadata description – in particular the refinement 'quality report'.

The element 'description' itself should consist of brief and clearly structured textual comments or abstracts which charaterize the dataset content, type of resource, its aims and background and expected quality. The description element is often used by users to assist their selection of appropriate resources from a set of search results. If such information is not available to the data provider for information resources, the compilation may require considerable effort.

The newly added refinement 'quality report' required additional time input to identify and provide the necessary information. The currently proposed structure of the 'quality report' was regarded as a useful platform for data providers to make crucial information accessible to data users. Thus none of the partners regarded any of the suggested headings for the quality report as 'not important' and is thus not represented in Figure 23. Each of the topics was considered interesting to very important. Provision of definitions was seen as the central aspect to be included within the quality report. Furthermore, it was confirmed that the quality report is well placed as a refinement under the element DCMI element description. However clear guidance should be provided on how to best structure a quality report for different data types.

Although discussions concerning the quality aspect were vigorous and extensive, the current proposed form of the element refinement 'quality report' was seen as a step forward and a useful addition to the metadata schema.

The issue of accessibility and rights was a cause of further lively discussion by the NEFIS data providers.

The DC element 'rights' as presented in the NEFIS metadata schema allowed users to select from the options: (a) Public access – no limitations; (b) limited access – general restrictions, unspecified; (c) restricted access – need to purchase; (d) restricted access – need to be a member; and (e) restricted access – need to register as a user. Partners considered their provided datasets as 'public', and few of them considered the option 'limited' to be relevant. There was a trend to attach some restrictions to public data use. "*Datasets should only be available and accessible for public use with the developed visualization and analysis tools, and should not be available for direct downloading by users*" (Figure 24).

It should be noted that the NEFIS datasets consisted mainly of aggregated data originating from complex statistics which may be available either freely through official publications or data services. If we assume that data users would be interested in accessing more detailed or even raw data, the responses of the data providers will most likely be different. Half of the data providers stated that a user would have to directly apply to an organization to receive more detailed or raw data. They would then make data available either free of charge or at a processing cost

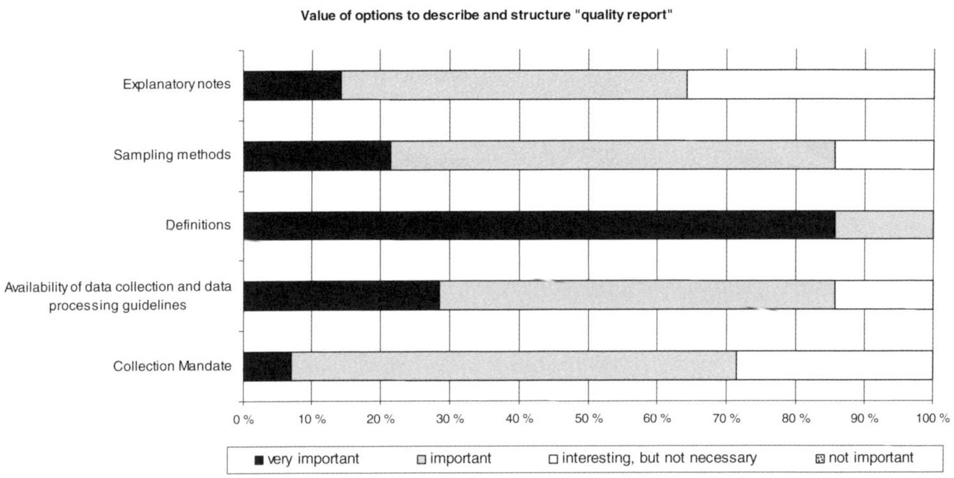

Figure 23. Importance of elements for describing the quality of an information resource.

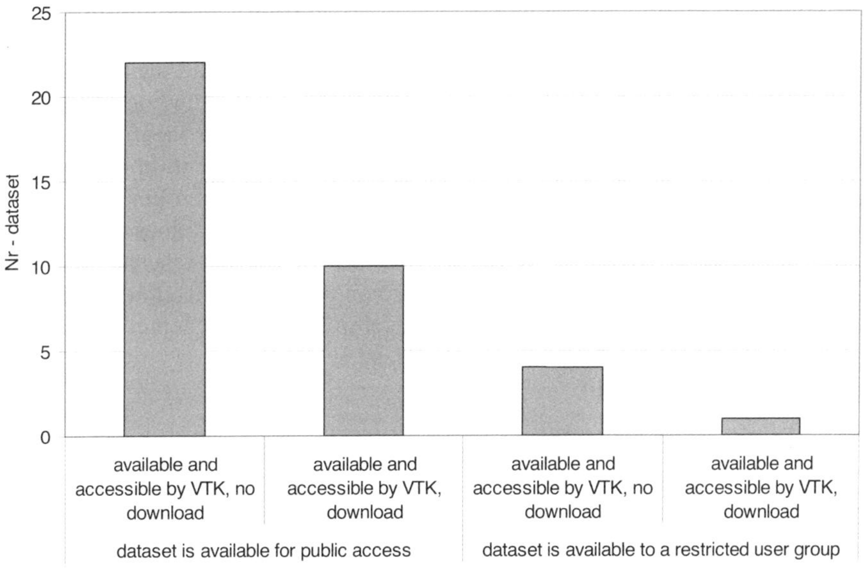

Figure 24. Forms of data access (VTK = visualization toolkit).

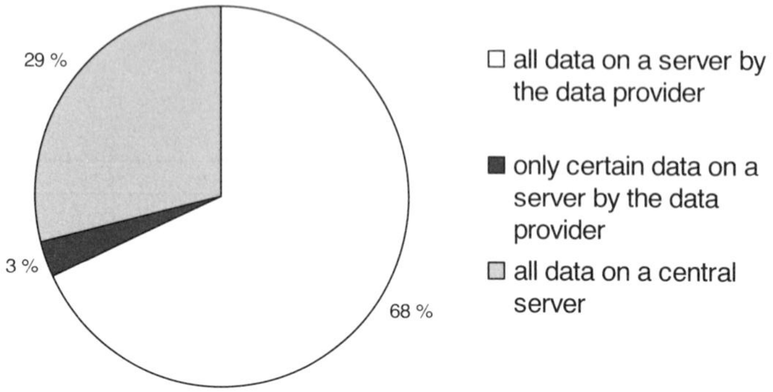

Figure 25. Preferred form of data storage.

(usually linked to certain conditions of data use). Nearly half of the respondents stated that no further data or raw data would be accessible in any form.

Further, the data providers expressed their opinions on responsiblities towards data access management. For most provided datasets the data provider should be responsible for data access management. This implies that a system needs to be based on a distributed network design – i.e. that data would be hosted on a server at the data provider's premises. The advantage of such an approach is that it caters for easier data management and data access. This was underlined in the survey as nearly three-quarters of data providers indicated that they would favour data to stay on their own servers (Figure 25). Those preferring the solution of a central server were often organizations who do not have the technical capacity to provide their data through on-line databases or Web services.

The overall conclusion from the evaluation was that the NEFIS metadata schema is applicable and usable in its current form, despite some existing obstacles. Both from the perspective of dataset documentation, but also for identification and retrieval of data resource, the schema was evaluated as sufficient. In particular for periodic reporting operations the schema can be considered a key asset. Clear dataset description and effective dataset retrieval are applications relevant and interlinked with such reporting processes. The NEFIS metadata schema fulfils these demands as its structure and content are harmonized. It can therefore be regarded as a milestone within the process of developing and implementing a fully operational European forest information system.

3.6 Interoperability of NEFIS datasets

The aims of the NEFIS project included the following: (1) making a small number of datasets accessible through a resource discovery component, and (2) making analysis of the datasets possible through a single visualization toolkit. Although these aims are relatively modest, they still highlight many of the problems which need to be addressed in terms of technical, semantic and organizational interoperability. These are significant for esablishing a European forest information system, and could not be fully resolved within the timeframe of the project. Nonetheless, fruitful lines of development have become clearer and a good start has been made in identifying the pressing issues to be resolved.

As part of the investigation of the dataset interoperability, NEFIS project partners were asked to provide templates of their dataset structures, to allow identification of common features, peculiarities and authorities in use, and thus establish what degree of interoperability might be achieved (note that NEFIS datasets are listed in section 4.3.2, Table 12).

Ten sets of sample dataset structures were reviewed, mostly detailing table structures with column and row headings. Of approximately 1400 terms listed, about 90% were unique, and those that occurred in more than one dataset did so almost exclusively only in other datasets from the same provider. To some extent this reflects the disparate nature of the dataset subjects, but it is clear that providers are not talking the same language, literally and semantically; and that datasets have not been designed with interoperability in mind. This is to be expected, and even when interoperability is recognized as desirable, considerations of backwards compatibility may have justifiably over-ridden thoughts of redesign.

Rather than attempt to manipulate these data structures into something for which they were not designed, it seems more appropriate to broaden the examination to a discussion of the issues which need to be addressed if useful interoperability is to be achieved within the forest sector. Therefore an attempt was made to identify the broader context in which this aim should be considered. Significant considerations in planning the route ahead were sought, without constraining the debate to particular solutions.

It is clear that the sample datasets submitted for NEFIS are not interoperable as they stand. To make them so would require, as a minimum:

At the dataset level:

- Adoption of an agreed controlled vocabulary for labeling variables (column and row headers) and for the data content where appropriate;
- Mapping existing labels to the identified standard;
- Identification and flagging of data which are not consistent with other datasets but could be made so by the application of a conversion factor;
- Creation of tables of those conversion factors;
- Identification and flagging of 'non-standard' data which cannot be made consistent;
- Identification or creation of reference documents explaining how, when and why the data were collected;
- The implementation of all the above in all required languages.

At the information system level:

- Definition of requirements for search interoperability, and the 'minimum kit' required for successful retrieval (which may differ for different end users);
- Definition of required structures for extended metadata;
- Creation of services to hold and maintain the agreed authority lists;
- Creation of some form of central catalogue;
- And, inevitably, much more work related to metadata.

The descriptive metadata so far compiled for NEFIS datasets is sufficient for the purposes of cataloguing and identifying relevant datasets using the resource discovery component of the currently available forest information system and general search engines, and to provide basic human-readable information on the content of those datasets with some indication of the standards used (if any). However, the metadata is not sufficient for the purpose of cross-searching selected datasets and compilation of combined data for submission to processing tools such as the visualization toolkit, without considerable user intervention to manipulate the data. This requires specialist knowledge and computer literacy that only part of the target audience of the advanced NEFIS demonstrator can be expected to have.

The 'top-level' or generic metadata as produced for NEFIS describes a dataset as an object. Interoperable searching requires more information about the data itself, including:

- Defined protocol(s) to translate queries between the enquirer's search client and the dataset server;
- Mapping of dataset field labels and commands to those understood by the protocol;
- Information on the field contents and any authority lists used to control the content;
- Sources of background information of which the enquirer should be made aware (for example, in making judgements of quality);
- Location and provenance of all associated files, and how and when they were created and stored;
- Data-specific intellectual property rights, if different from the dataset as a whole;
- Executables required to read or manipulate the data;
- Language and character set;
- Access control mechanisms needed to unlock secure data for permitted users.

4. The Advanced Information System Demonstrator

4.1 Objectives

As described in section 1.3.2 the system which was built as a result of the EFIS project (contract no. 17186-2000-12 F1ED ISP FI) during 2000–2002 is comprised of a metabase – or resource discovery component, and the data analysis and visualization tool which allows users to further process retrieved data within an Internet environment. The main advances in systems development within NEFIS were the specification of a more comprehensive and detailed metadata schema. The new schema was applied to the resource discovery component of the advanced NEFIS demonstrator. Secondly an additional set of analysis and visualization features were added in order to allow the various datasets of the NEFIS project to be explored to their full extent.

The main development, however, was embedding the actual operation of the information system to run within a network of distributed datasets. This means that the datasets are hosted by individual data providers and can be retrieved by a user from those locations. The advanced NEFIS demonstrator distinguished two approaches for data retrieval (see Figure 26 in which the upper figure represents approach 1, and the lower figure represents approach 2):

1. Retrieval of an individual dataset from one remote server location: data physically located on the data server of a NEFIS partner e.g. Inventaire Forestier National of France.
2. Retrieval of data (in our case forest inventory data for land and forest area) from multiple countries located on remote servers with one query request. This means that the data are retrieved from physically different locations (servers) simultaneously and compiled into one table.

Both approaches allow further exploration and visualization of the data once they have been retrieved. In the first approach there is no direct comparison of the data from different countries. The second approach, presented in Figure 26 (approach 2), is combining datasets (in tables, maps and graphs) that may be based on different definitions of forest, and are therefore not directly comparable. Harmonization of actual data from different countries, in our case national forest inventory data, was not part of the NEFIS project, but is subject to other activities such as the FAO Forest Resources Assessement process (FAO, 2004) or the COST Action E43 on 'Harmonisation of National Forest Inventories in Europe: Techniques for Common Reporting' (COSTE43, 2004). The aim within NEFIS has been to show that it is technically possible to perform multi-host data searches and retrieval. The project has not contributed to the issue of data harmonization.

The following sections will present the main elements of the advanced demonstrator – (1) the resource discovery (RD – section 4.2); (2) the visualization toolkit (VTK – section 4.3); and (3) describe in more detail the remote search demonstrator (section 4.4 – retrieval of data from multiple servers by placing a single query request).

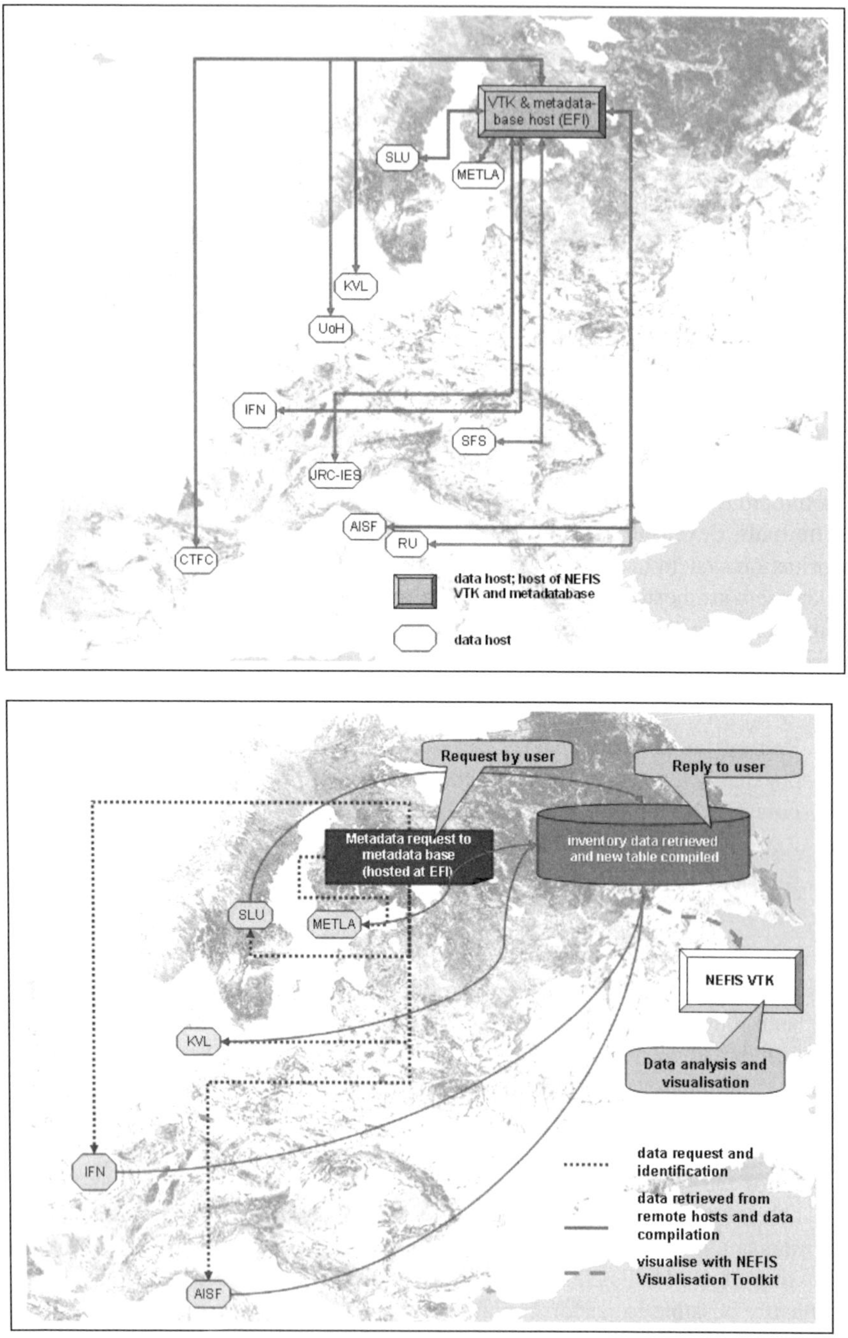

Figure 26. Remote data retrieval from data providers with the advanced NEFIS demonstrator. The upper figure (approach 1) represents the retrieval of one individual dataset from a remote host. The lower image shows simultaneous retrieval from multiple hosts (approach 2).

4.2 The resource discovery component

The RD component allows the data holder to provide metadata, based on the NEFIS metadata schema, and the user to retrieve metadata records and data as requested in a query. The interface to implement these actions is web-based and constructed in a similar way to conventional search engines. The metadata comprehensively describes an information resource and may include either a direct link to the web location where the dataset is hosted, or the ability to retrieve data from distributed data servers (Figure 26, section 4.1). The user has the option to visit the identified resource and/or directly request the data. In order to better understand the design and functioning of the RD component a number of terms used in the RD context are presented at this point:

- **eXtensible Markup Language** (XML) – a meta language used to define domain specific markup languages.
- **interoperability** – the ability of systems to operate together.
- **metabase** – a database containing metadata – also referred to as "metadata database".
- **information resource discovery** – the task of finding information (e.g. data tables) or information resources (e.g. libraries).
- **schema** – a complete description of the boundary conditions of a system, for example a metadata definition or a data table.
- **structured data** – data that can readily be described by a schema, for example a database, as opposed to unstructured data such a Word document containing a complex collection of arbitary data types.

The metadata software for the RD component was first developed for the EFIS project between 2000 and 2002, and then further extended to meet the needs of the NEFIS project. In order to cater for such development, the metadata system software was designed as a flexible and extensible system based on a XML schema that comprehensively describes each of the metadata elements and which is used to dynamically build the system's web forms. The metadata content, populated via the forms by users, is then stored as XML elements in a relational database. The advantage of this approach is that the metadata content can be enriched without having to change the database design (Figure 27).

The RD component is designed as a cataloguing system which appears as a single resource discovery metadata layer to the user. Beneath the resource discovery layer that is presented to the user a number of further layers of metadata must be considered. The following layers of metadata within the RD component can be identified:

- definition of the metadata elements – the XML metadata schema (Layer 1)
- the content of the metadata elements – the forest content metadata (Layer 2)
- the data schema defined by the services provided by the Remote Search Demonstrator (Layer 3)

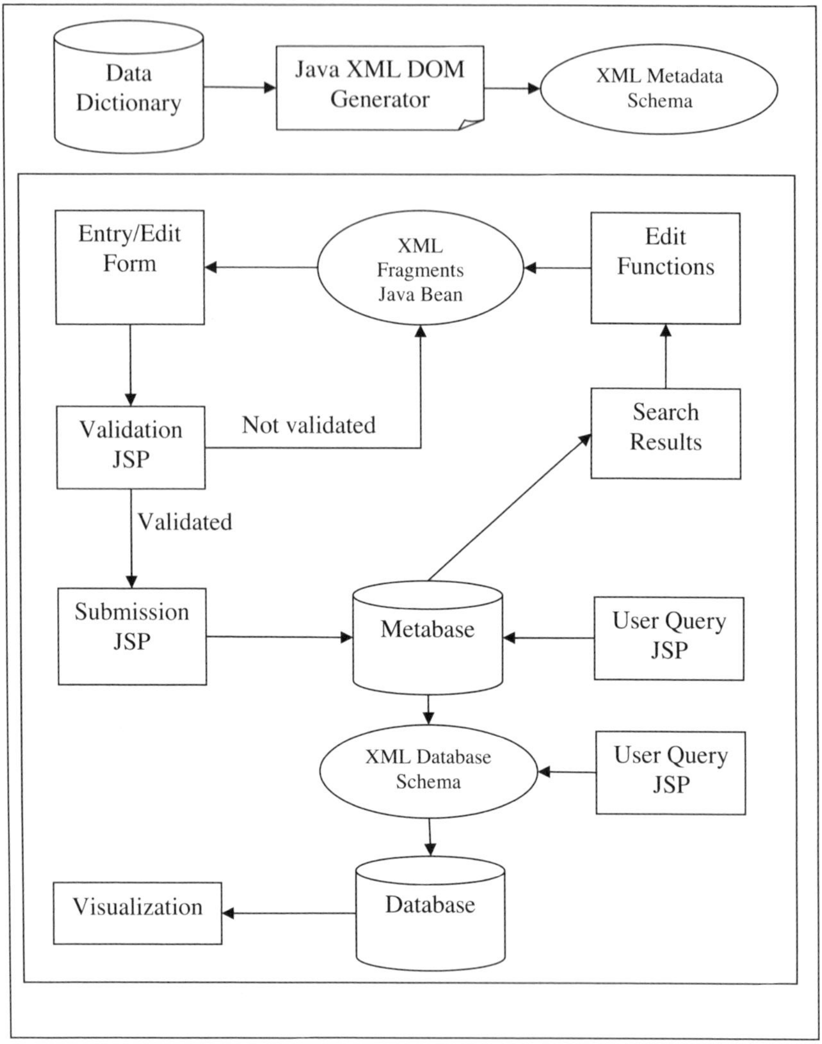

Figure 27. The functional components of the metadata system (JSP = JavaServer Pages).

Firstly, the metadata elements are defined and described that will be used in the catalogue. This description is a comprehensive description of each metadata element – its name, registration authority, data type, etc. It contains no content values – only the description of the elements (Layer 1; see Figure 28). Populating these elements with forest information content then constitutes the metabase or metadata catalogue (Layer 2). Having populated the catalogue one can now find (or discover) information resources and, where appropriate, link to them. The references may be to either structured or unstructured information resources. In

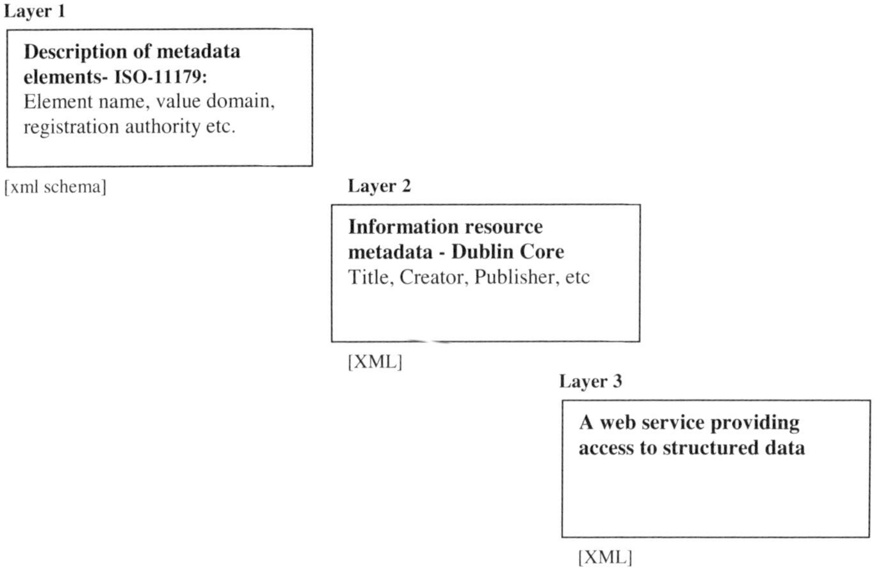

Figure 28. The metadata hierarchy.

the case of structured information resources, such as databases or data tables, there is another layer of metadata that describes that structure, the data schema. If this structure (e.g. database or data table structure) is recorded, then there is a mechanism available to go into information resources and directly extract data from them (Layer 3).

Layer 1 is described using a subset of the ISO11179 standard for the description of data elements (ISO, 2005). The DCMI metadata element set, on which the NEFIS metadata schema is based, is used to capture metadata on forest information resources (Layer 2). Lastly, Layer 3 is used to describe structured information services that may be distributed as a family of Web services. Further detail on each of these metadata layers is provided in the following sections.

4.2.1 Layer 1 – The metadata dictionary

The metadata definition for Layer 1 is encapsulated in a schema. The schema stores each of the elements, their definition, data types, attributes and value domains. The schema, therefore, acts as the definition of the metadata.

4.2.1.1 The XML expression of the metadata schema (metadata elements)

Each of the metadata elements is described using attributes and elements expressed in XML. A generic XML structure is used to describe each element:

```
<MDElement attribute=value>
        <Name></Name>
        <Comment></Comment>
        <Domain></Domain>
        <ControlledValues></ControlledValues>
        <Qualifier></Qualifier>
</MDElement>
```

The ISO 11179 standard for the description of data elements is used to describe the individual DC metadata elements. In addition to ISO 11179 there have been added value domain ranges and, where applicable, controlled values in order to fully describe each element and qualifier (element refinement or encoding scheme). There are 10 element attributes including one attribute for comments and 3 describing the content of an element (Box 4).

All the DC metadata elements have attributes of unlimited maximum occurrence and so-called 'string' data type. Both of these attributes are explicitly specified in the metadata schema. For example the unqualified element 'creator' is expressed in XML in Box 5:

Box 4. Element Attributes

1. Name - The name of the data element.
2. Identifier - The unique identifier assigned to the data element.
3. Version - The version of the data element.
4. Registration authority - The entity authorized to register the data element.
5. Language - The language in which the data element is specified.
6. Definition - A statement that clearly represents the concept and essential nature of the data element.
7. Obligation - Indicates if the data element is required to always or sometimes be present - true or false
8. Datatype - Indicates the type of data that can be represented in the value of the data element - see list of data types in the Data Types section.
9. Maximum occurrence - Indicates any limit to the repeatability of the data element. 0=none, -1 = no limit.
10. Comment - A remark concerning the application of the data element.

Element Content:
1. Value - The value of the element in the specified Datatype.
2. Domain - The range of permitted values.

Controlled values - a list of controlled values - such as country names.

Box 5. Unqualified element 'creator'

= <MDElement ControlType="**text**" DataType="**string**" Definition="**An entity primarily responsible for making the content of the resource.**" Identifier="**Creator**" Language="**eng**" MaximumOccurrence="**3.0.**" Obligation="**true**" RegistrationAuthority="**DCMI**" Version="**1.1**">
<Name>**Creator**</Name>
<Comment>**Examples of a Creator include a person, an organisation, or a service. Typically, the name of a Creator should be used to indicate the entity.**</Comment>
<Domain>**null**</Domain>
<ControlledValues>null</ControlledValues>
</MDElement>

The example in Box 5 represents an 'unqualified element'. A qualified element as compared to an unqualified element may contain extensions to the particular DC metadata element. Qualifications can be distinguished by their registration authority attribute. For example, it can show qualifications as developed within the projects EFIS or NEFIS thus using a different registration authority (i.e. 'EFIS' or 'NEFIS'). An element added for NEFIS was 'reference system' and would therefore read as RegistrationAuthority="NEFIS".

Qualifiers that have been adopted outside given standards can be regarded as highly useful additions to a metadata schema. They can be implemented as sub-elements inside the main element. To do this an element is qualified by adding one or more additional 'qualifier sub-elements' within the main element. The example in Box 6 extends the DCMES 'creator' element as used above with two additional qualifiers – 'creator address' and 'creator type'. Within 'creator address' there are in additional divisions such as (1) street address, (2) city, (3) post code, etc.

It is evident from the above example that the expression of the elements and sub-elements can become very verbose. However, all of this information can be machine processed, although if necessary also read and understood by humans.

Box 6. Qualified element 'creator' with sub-element "address".

<MDElement ControlType="**text**" DataType="**string**" Definition="**An entity primarily responsible for making the content of the resource.**" Identifier="**Creator**" Language="**eng**" MaximumOccurrence="**3.0.**" Obligation="**true**" RegistrationAuthority="**DCMI**" Version="**1.1**">

<Name>**Creator**</Name>

<Comment>**Examples of a Creator include a person, an organisation, or a service. Typically, the name of a Creator should be used to indicate the entity.**</Comment>

<Domain>**null**</Domain>

<ControlledValues>**null**</ControlledValues> = <Qualifier Identifier="**Address**" RegistrationAuthority="EFIS" Version="1.1">
Address

<StreetAddress Comment="Street address." ControlledValues="**null**" DataType="**string**" Definition="**The street address of an address**" Domain="**null**" Identifier="**Address**" Language="**eng**" MaximumOccurrence="**2.0**" Obligation="**false**" RegistrationAuthority="**EFIS**" Version="**1.1**">**Street Address**</StreetAddress>

<City Comment="**The city element of an address**" DataType="**string**" Definition="**The city element of an address**" Domain="**null**" Identifier="**City**" Language="**eng**" MaximumOccurrence="**1.0**" Obligation="**false**" RegistrationAuthority="**EFIS**" Version="**1.1**">**City**</City>

<Region Comment="**The region element of an address**" DataType="**string**" Definition="**The region element of an address**" Domain="**null**" Identifier="**Region**" Language="**eng**" MaximumOccurrence="**1.0**" Obligation="**false**" RegistrationAuthority="**EFIS**"Version="**1.1**">**Region**</Region>

<PostCode Comment="**The postcode element of an address**" DataType="**string**" Definition="**The postcode element of an address**" Domain="**null**" Identifier="**Postcode**" Language="**eng**" MaximumOccurrence="**1.0**" Obligation="**false**" RegistrationAuthority ="**EFIS**" Version="**1.1**">**Post Code**</PostCode>

<Country Comment="**The country element of an address**" DataType="**string**" Definition="**The country element of an address**" Domain="**null**" Identifier="**Country**" Language="**eng**" MaximumOccurrence="**1.0**" Obligation="**true**" RegistrationAuthority="**EFIS**" Version="**1.1**">**Country**</Country>

<Telephone Comment="**The telephone number element of an address**" DataType="**int**" Definition="**The telephone number element of an address**" Domain="**null**" Identifier="**Telephone**" Language="**eng**" MaximumOccurrence="**1.0**" Obligation="**false**" RegistrationAuthority="**EFIS**" Version="**1.1**">**Telephone**</Telephone>

<Fax Comment="**The fax number element of an address**" DataType="**int**" Definition="**The fax number element of an address**" Domain="**null**" Identifier="**Fax**" Language="**eng**" MaximumOccurrence="**1.0**" Obligation="**false**" RegistrationAuthority="**EFIS**" Version="**1.1**">**Fax**</Fax>

<Email Comment="**The email number element of an address**" DataType="**string**" Definition="**The email number element of an address**" Domain="**null**" Identifier="**Email**" Language="**eng**" MaximumOccurrence="**1.0**" Obligation="**false**" RegistrationAuthority="**EFIS**" Version="**1.1**">**Email**</Email>

Box 6. Continued.

<WebPage Comment="**The URL element of an address**" DataType="**string**"
Definition="**The URL element of an address**" Domain="**null**"
Identifier="**Web page (URL)**" Language="**eng**"
MaximumOccurrence="**1.0**" Obligation="**false**"
RegistrationAuthority="**EFIS**" Version="**1.1**">**Web Page**</WebPage>
</Qualifier>
= <Qualifier ControlType="**select**" DataType="**string**" Definition="**The type of Creator organisation**" Identifier="**CreatorType**" Language="**eng**"
MaximumOccurrence="**3.0.**" Obligation="**true**"
RegistrationAuthority="**EFIS**" Version="**1.1**">
Creator Type
<CreatorType>Creator Type</CreatorType>
</Qualifier>
</MDElement>

4.2.1.2 NEFIS metadata entry and editing system

The actual metadata entry and editing system is comprised of a set of Java Server Pages (JSPs) that read the XML schema, parse it and convert it into a web form. The example that was presented as an XML schema of the qualified 'Creator' element is converted to JSPs and appears seen through a browser as presented in Figure 29.

Titles of mandatory elements are highlighted in red (grey in Figure 29) indicating that this information must be provided in order to successfully upload a particular metadata record to the metabase. The qualifiers of 'Creator' are 'Address' and 'Creator Type' and are not mandatory. It can be seen in the XML schema above that 'Address' is a complex qualifier made up of numerous further divisions whereas 'Creator Type' has no further divisions.

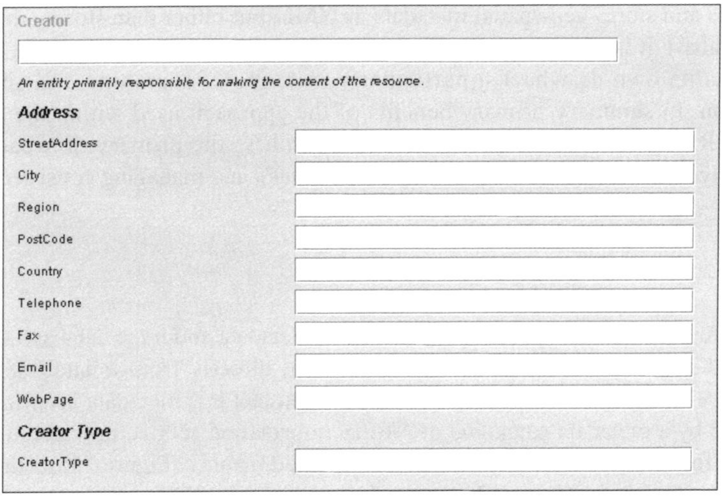

Figure 29. JSP of the Metadata Entry form for the DCMI Element 'Creator'.

4.2.2 Layer 2 – The metabase

The metabase refers to the data repository in which the metadata content is stored; currently this is a relational database. Once a metadata provider has entered metadata content to the RD entry form this content is stored in the metabase. This constitutes layer 2 of the metadata in Figure 28.

Because the XML metadata schema is extensible it can be changed and those changes will be reflected in the rendition of the web forms. For example a second street address might be required. In this case an additional sub element is added to the XML schema and this would appear in the web pages. Such a change may mean that the database design of the metabase will need updating. As the RD component is designed such that the metadata content is encapsulated in XML no change to the metabase design is required. The database design only needs to be modified if additional elements are required. No change is needed for sub-elements.

Therefore, the metabase fields in the current system store fragments of XML that encapsulate all content relating to that element, including all qualifications and possibly repeated elements. The described XML schema defines a tree structure that has metadata as the trunk, metadata elements as the branches and qualifications as the twigs and leaves. The metadata provider populates this tree structure with content and gives values for the branches, twigs and leaves. The branches of the tree, which are equivalent to a metadata element, are then cut off, together with all of their twigs and leaves, and deposited as XML into the metabase as a character string. This approach results in a very simple database, but very rich content. Searching can be performed on a metadata element basis. The approach does have the drawback of being very wordy, with a considerable amount of data redundancy, when compared to a normalized relational database. However, what is lost through redundancy is made up for through flexibility and extensibility as the schema may be enriched without the need to make changes to the database.

There are of course tradeoffs to taking this approach. It was opted to store the branches of the XML in a database. A similar system is ESRI's ArcCatalogue which compiles and stores geo-spatial metadata as XML, but rather than storing the XML in a database it is stored in individual XML files – one for each metadata record. This has its own drawbacks, particularly in terms of searching and speed of operation. In summary primary benefits of the approach used within the current NEFIS RD system are flexibility and extensibility; the primary drawbacks are storage (could be reduced by optimizing tag names), and managing consistency.

4.2.3 Layer 3 – service layer

In the previous EFIS project, Layer 3 provided metadata about the database schema, encoded in XML. This enabled data to be read directly from a database server hosted by a single data provider. In the NEFIS project this metadata layer has been replaced by a metadata catalogue of NEFIS information service providers that can provide forest information services in a distributed manner (Figure 26, section 4.1). For example, a Layer 2 metadata record can refer to a forest information service that is provided not by a single data provider, but by a family of data providers across Europe and whose collective data comprise the information service. The single information service therefore consists of data from many information providers. The information service collates the results generated from the family of information service providers. The metadata to describe the family of providers is held in a simple XML file that catalogues the providers details. Each provider must host a simple Remote Procedure Call server (RPC-server) that provides a standard set of information services.

Having found a description of an information service using the Layer 2 (resource discovery) metadata, the user can access distributed information with the aid of Layer 3 metadata, to collect and collate information from the service. Note that Layer 3 metadata should be invisible to the user, who probably has little interest in accessing it. Rather the user is interested in accessing the service enabled by this layer of metadata. Having located and collated the information from the distributed family of information providers the user is then able to visualize the information using the VTK.

4.3 The visualization toolkit: what is it and why use it?

4.3.1 Exploratory data analysis

An operational forest information system needs to allow seamless discovery of information from other existing forest information systems or databases, and the retrieval and exploration of data from such sources. The development of a common metadata model caters for a harmonized access to information or actual data. The discovered resource(s) (data tables/databases) from one or more sources may then be available for downloading and processing and further visualized and investigated using a flexible data explorer. In this way the data can easily and seamlessly be further utilized in a real time environment from any given location with access to a Web browser. The VTK was first designed for the EFIS project and then further developed within NEFIS. The VTK itself has been built on the basis of the CommonGIS system for thematic mapping and exploratory data analysis (Andrienko et al., 2003). The VTK can be used in an Internet environment allowing a user to get quickly familiar with retrieved data, judge their usability for a particular purpose and ultimately further explore and visualize the data.

The capabilities of the VTK reach far beyond the mere generation of maps and graphs. Particularly, the VTK can be used for a comprehensive exploration of new, previously unknown data. Exploratory Data Analysis, or EDA (Tukey, 1977) is a philosophy and discipline of unbiased examination of data aimed at detecting and describing patterns, trends, and relationships in the data rather then getting answers to particular questions or checking some pre-conceived hypotheses. In other words, an explorer approaches a dataset open-mindedly being willing to perceive what the data might tell him/her rather than scanning the data for particular aspects and features in an attempt to verify already existing ideas and opinions.

The role of EDA in NEFIS was to help users to get acquainted with new data. Since in NEFIS forest-related data are to be provided to people through the Web, these people need tools to explore the data they find. The VTK can be useful not only for the 'external world', but also for the forest experts themselves, in particular, when they have to analyze new, unfamiliar data.

4.3.1.1. General principles of data exploration

Typically, any data exploration starts with an attempt to overview the entire dataset and to grasp its major distinctive features (Andrienko and Andrienko, 2006). Often some characteristics of the data preclude an overall view of the entire dataset:

- very large data volumes; and
- multidimensionality: the data refer simultaneously to two-dimensional geographical space, and to time.

Such problems occur quite often in data exploration, and there are standard approaches to deal with them. The problem of very large data volumes is solved by means of data aggregation. There are some 'rules of thumb' concerning aggregation:

- In computing characteristics of aggregates from the individual data values, be cautious with averaging: pay attention to the value range, character of the statistical distribution of the values, and presence of outliers.
- Prefer positional statistical measures (i.e. medians, quartiles, percentiles, etc.) to means.
- Do not rely upon a single aggregation; vary the level and method of aggregation.

However, even the best aggregation involves significant information loss. Therefore, aggregation should be complemented with exploration of individual data instances, in particular, global and local outliers and other sorts of anomalies such as "strange" behaviours or unusual value combinations. Examination of various "particulars" typically occurs after the exploration into general patterns, trends, and regularities.

The problem of multidimensionality is approached by means of "slicing" the data. For data with two dimensions, space and time, there are two ways of slicing:

- Consider the spatial distribution at different time moments (and, accordingly, the evolution of the spatial patterns over time).
- Consider the temporal behaviours at different locations in the space (and, accordingly, the variation of the behaviours over the space).

Each method of slicing provides a different perspective of the data and, hence, both methods should be used to complement each other.

For looking at the spatial distribution and its evolution over time, it is useful to apply spatial aggregation of the data; specifically, aggregation of the data by cells of a regular two-dimensional grid. For having an overall view of the whole multitude of the temporal behaviours, two methods of aggregation on a time graph display are applicable: (i) computation of positional statistical measures (e.g. deciles); and (ii) counting values fitting in specified intervals.

In the absence of general spatial and temporal patterns and trends, focusing on small data subsets and individual instances becomes very important (actually, investigation of various "particulars" is necessary even when general patterns exist – in this case, at least deviations from such patterns require attention). A feasible approach in a case of large data volumes is to focus on behaviours with specific characteristics, for example, with sudden rises and/or drops of tree defoliation. Such behaviours can be extracted from the mass by means of querying, in particular, subset selection through interaction with a display. When some behaviour of interest is located on a map, it is appropriate to look at the behaviours in the neighbouring locations: are they similar or completely different?

4.3.1.2. Tools and techniques

As mentioned in section 4.3.1.1, it is necessary to look at the data on different levels: (i) overall (entire dataset); (ii) intermediate (various subsets); and (iii) elementary (individual instances). Clearly, "looking" at the data essentially requires visualization. For looking at the spatial distribution, the primary visualization tool are maps. Time graph tools support investigation of temporal behaviours.

For dealing with the large data volume, aggregation tools are used: for example, spatial aggregation by cells of a regular grid and aggregation according to attribute values on a time graph. Individual data instances and data subsets for a more detailed exploration can be selected by means of querying tools. There are two modes of querying:

- Filtering: only data satisfying some specified constraints are shown; the remaining data are hidden.
- Marking: graphical items corresponding to data of interest are specially marked (highlighted) among the other items.

One can also look at options for dynamically linking between different data displays. Through dynamic linking it may be possible to select behaviours on a time graph and see on a map where they are in space. And conversely, a user may select locations on a map and see the corresponding behaviours on a time graph. In addition to the linking by simultaneous marking of corresponding display items, displays can be linked through propagating class colours from a map.

Besides aggregation and querying, certain techniques of data transformation are further options for data exploration.

- Standard normal transformation or standardizing and bringing together originally incomparable data (for example, referring to different species).
- Temporal smoothing for disregarding minor fluctuations and concentrating on major temporal trends.
- Computing differences with regard to a specific time moment (year) for disregarding absolute value differences and looking for similarity of behaviour trends in different places.

- Computing differences with regard to the previous time moments for comparison of behaviour trends in different locations and for detecting behaviours with extreme increases or decreases of values.

In summary, any serious data exploration requires a good integration and coordination of a variety of complementary tools which:

- support both overall and detailed views;
- look at data from different perspectives and explore different aspects;
- overcome difficulties such as large data volumes and multidimensionality;
- discover regularities and detect various anomalies;
- find data of specific interest and consider all potentially relevant information.

4.3.2 Data types

Due to the nature of the partnership in NEFIS, a wide variety of datasets were made available. They range from statistical data from national forest inventories, socio-economic data, timber prices and forest products trade data, to spatial information on forest fires, land use and rural development. Table 12 lists the data types and their geographic coverage which varied from global (e.g. forest products trade and trade flows) to the level of regions within a particular country (e.g. forest fires in the Umbria region of Italy). NEFIS intended to demonstrate not only the various types of data which exist on forests, but also take into consideration, spatial and temporal aspects as well as differences in data formats with which an operational European forest information system will be confronted.

The VTK therefore needed to prove its applicability to this broad array of datasets and their needs for differing features for exploratory data analysis and visualization. In the following sections the basic prinicples of EDA will be further underlined by presenting a set of examples based on the NEFIS data.

It should be noted that the VTK is just one example of a data analysis tool – in the case of the NEFIS project embedded within a forest information system. For the NEFIS demonstration, the datasets had to be prepared specifically for use with the VTK. If standards for the harmonized metadata descriptions of dataset structures were developed and detailed metadata descriptions of dataset structures became available, then tools such as the VTK could be developed in such a way that the need for human intervention is lessened, or even removed.

Table 12. NEFIS datasets.

International data	Coverage
Temperate and Boreal Forest Resources Assessment 2000	country level
Long-term forest resources assessment database	country level, time series
EU socio-economic data	country/regional level
Forest products trade flow database	country level, time series
Forest research capacities database	country level
ICP-Forest data: • Level 1 crown condition; soil data; foliage (1994–1997)	country level
Forestry experimental data (the NOLTFOX database)	country
Daily Fire risk maps	Europe, time series
Seasonal burnt-area map	Europe
European forest map	Europe (ArcView/ArcInfo comp.)
European eco-zone map	Europe (ArcView/ArcInfo comp.)

Country data	Coverage
National forest inventory statistics, France	regional level
National forest inventory statistics, Finland	regional level
National forest inventory statistics, Sweden	regional level, time series
National forest inventory statistics, Denmark	country/regional level
National forest inventory statistics, Italy	regional level
Aggregated national forest inventory statistics, Hungary	country level
National forest inventory maps database, France (metadata available only) National forest inventory photographs databases, France (metadata available only)	regional level
Roundwood price statistics, Finland	regional level, time series
Trade in secondary wooden products; removals and production, Hungary	country level
Various spatial datasets, Italy • Digital elevation model (75 m) • CORINE 4th thematic level • Climatic raster maps (month temperatures and rainfall) • Forest type maps • Rural development plans	country level (ArcView/ArcInfo)

Table 12. Continued

Regional data	Coverage
Forestry statistics: • forest map; forest inventory statistics; interventions made with Reg. 2080/92; fire risk; sub regional limits (mountain areas; protected areas, Natura 2000)	one region in detail (Umbria, Italy)
Land Cover Map of Catalonia	one region (Catalonia, Spain)
Rural development and land use: • Land use classification in Catalonia • Age structure (population) • Unemployment • Environmental and rural forest enterprises data (2001) • Migration • Landscape and ownership fragmentation • Economic support received per age and gender • Rural tourism infrastructures	one region (Catalonia, Spain)

4.3.3 Applications of the visualization toolkit – some examples

The major functionality and tools that were designed for the NEFIS project include a broad variety of options which were driven partly by the data types made available by the data providers:

1. Powerful tools for Internet mapping that support a variety of standard formats of map and table data.
2. A flexible client-server architecture that optimizes download time and supports integration of data from network-distributed servers.
3. A variety of interactive mapping techniques combined with statistical graphics displays and computations.
4. Comprehensive tools for analysis of spatial time-series, including animated maps, time-aware map visualizations, and statistical graphics displays.
5. Novel information visualization tools (dynamic query, table lens, parallel coordinate plots etc.) dynamically linked to maps and graphics via highlighting, selection, and brushing.
6. Possibility to complement interactive visual data analysis by mathematical methods of statistics and data mining.
7. Tools for interactive aggregation of grid data tightly coupled with dynamic visualization of aggregation.

Examples 1–14 show a variety of visualizations applied to the datasets collected during the project. Each visualization will be accompanied by an analysis task for which it can be effectively used.

Note that examples 1–14 are intended to illustrate the potential of exploratory data analysis within a forest information system. They are only snapshots as both data analysis and visualization are interactive processes. Some of the examples are shown here in colour for better illustration.

However, all examples are best viewed in colour. The colour snapshots as well as the instructions for recreating the visualizations using the VTK are given on the NEFIS web pages (www.efi.int search for nefis).

Example 1. Study spatial distribution with unclassified choropleth maps.

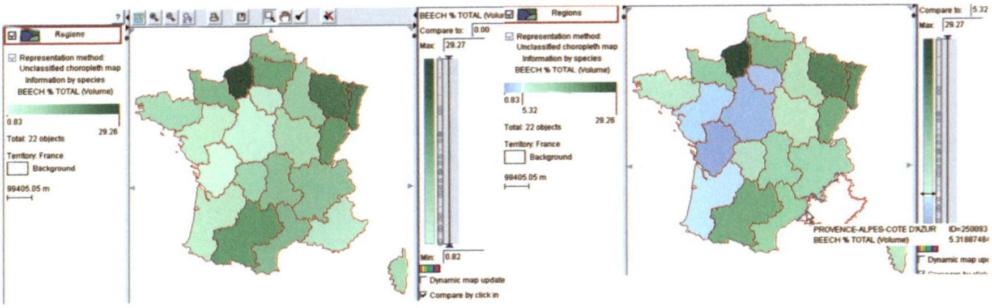

In Example 1, the first map shows proportions of beech (volume of timber) in forests within different regions of France. The second map provides a comparison of the region Provence-Alpes-Cote D'Azur (white polygon) with the other French regions. Regions with a higher proportion of beech volume than the chosen region are shown as green, and regions with a lower proportion than the chosen region are shown in blue. Using these techniques, one can study the general pattern of the spatial distribution and locate clusters of low and high values.

Example 2. Study spatial distribution with classified choropleth maps.

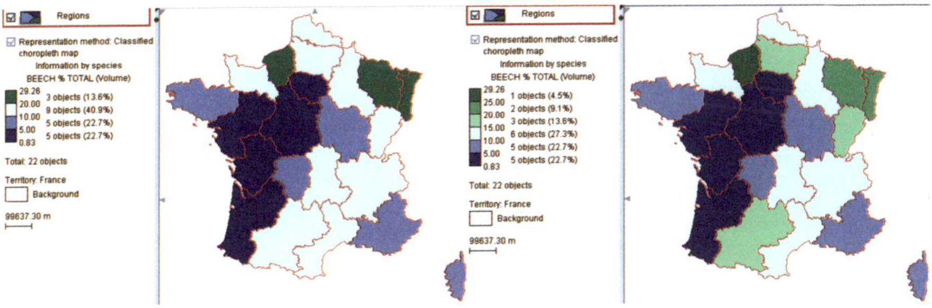

In the two maps of Example 2, the proportion of beech volume is shown as classified maps using two classification schemes (different level of detail in break specification: in the map on the left there are four classes; and in the map on the right there are six classes). Using this technique, one can construct a simplified and highly generalized image of the spatial distribution of beech. Users have the opportunity to change where the class breaks occur, and therefore fine-tune the visualization.

Example 3. Overview of spatial distribution of multivariate data.

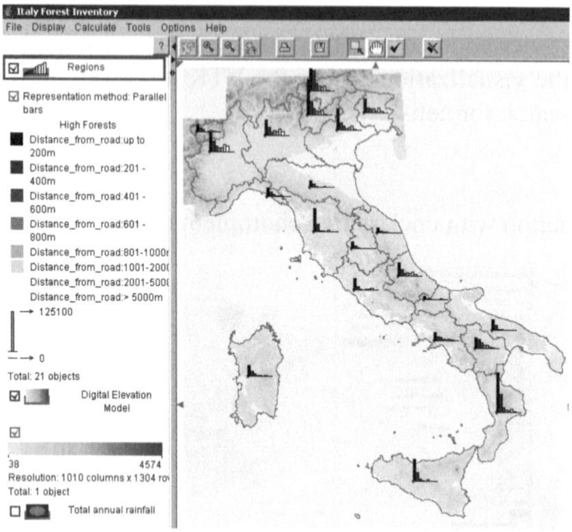

The bar charts in the map of Example 3 represent the area of high forests (i.e. crops or stands of trees, generally of seed or seedling origin, that normally develop a high closed canopy) at different distances from roads for regions of Italy. Using this visualization technique, the user can study differences in the accessibility to high forests and visually group similar structures between regions.

Example 4. Analysis of spatial distribution of multivariate data.

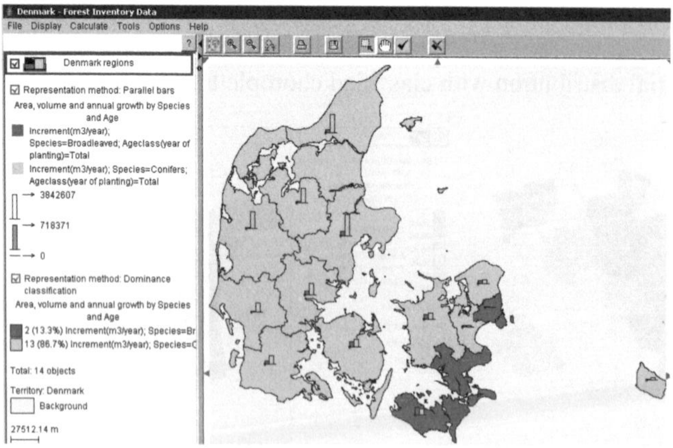

In Example 4, the bar charts which represent the increment of broadleaved and coniferous forest in Denmark (by regions) are enhanced by a complementary representation. The underlying dominant classification map shows which species has a larger increment. In this case, the darker regions in south-eastern Denmark show the regions where the increment in broadleaved forest is greater than the increment in coniferous forest. Using this visualization approach, the user can study the forest increment distribution in each region and identify regional groupings.

Example 5. Studying relations between attributes.

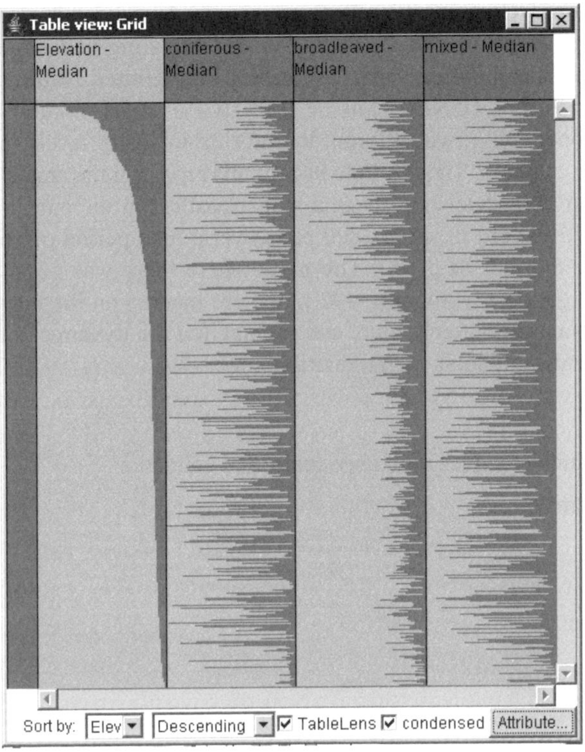

In the display shown in Example 5, the table rows are ordered according to the elevation values. The other columns show proportions of territories in Italy covered by coniferous, broadleaved, and mixed forests. Using this visualization, one can investigate if there are any correlations between values of several attributes.

Example 6. Studying relations between attributes using marking.

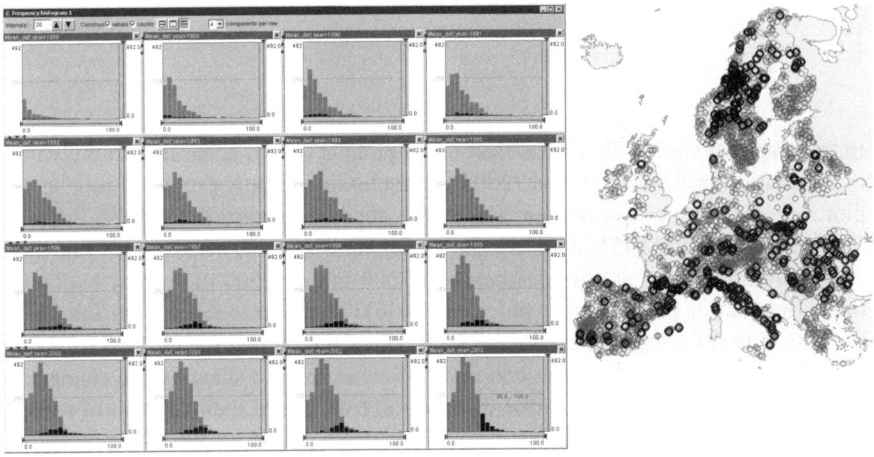

The data in Example 6 (and in Examples 7 and 8) originates from data collection activities of the International Co-operative Programme on Assessment and Monitoring of Air Pollution Effects on Forests (ICP-Forests) operating under the Convention on Long-Range Transboundary Air Pollution of the United Nations Economic Commission for Europe (UNECE) as of 1985. ICP-Forests monitors the forest condition in Europe, using two different monitoring intensity levels in cooperation with the European Union. The multiple histograms represent the annual mean defoliation per plot for all species present in a plot (classified in increments of 5% from 0% to 100% defoliation) in some 2800 plots over a time period of 16 years (1988–2003) (y-axis=number of plots). The plots where there was >35% defoliation in 2003 (lower rightmost histogram, 492 plots) are marked on the map (dark circles). Using this visualization technique, one can analyze the dynamics of values for selected objects and study their spatial distribution.

Example 7. Studying dynamics of values with user-controlled animation.

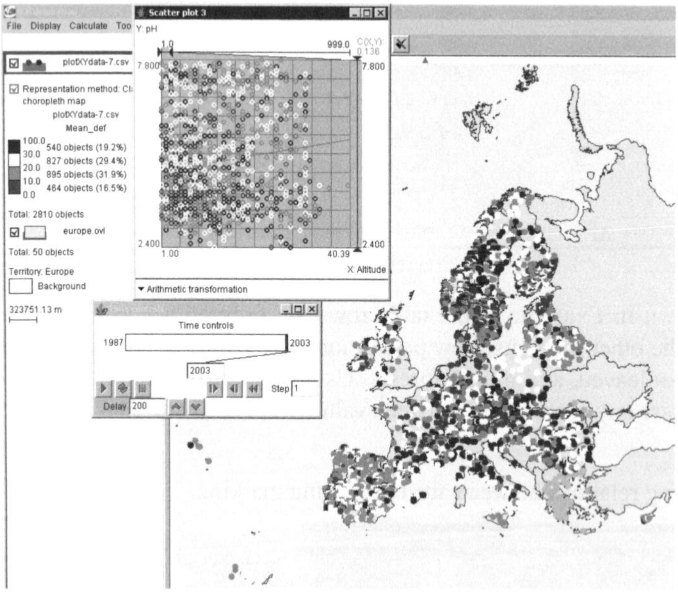

In the map in Example 7, a classification of plots is shown according to the values of defoliation in a selected year (2003). The classification is projected to a scatter plot representing soil properties. The x axis represents altitude in 50 m intervals (1=0–50 m and 40=1951–2000 m; the scale went up to 51 (≥2500 m), but there were only three plots at altitudes between 2000 m and 2500 m, and by excluding these the scatter plot is made more readable). The y-axis represents the pH of the soil. The different coloured circles represent the annual mean defoliation per plot. The illustration here is the end point of an animation showing the changes of defoliation values and their spatial distribution (map), and their relation to soil pH and altitude (scatter plot).

Example 8. Dynamic queries for relating multiple attributes.

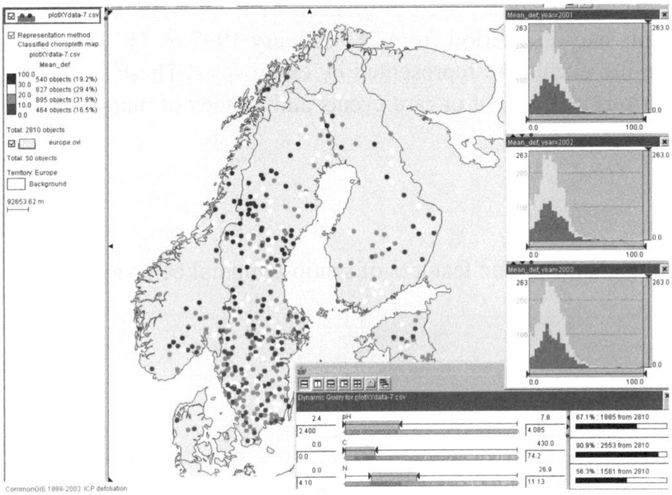

The classification map of Example 8 displays the defoliation in a given year; the histograms represent the distribution of the defoliation values in several years. The dynamic query tool (bottom right of the diagram) is used for selecting plots with particular soil characteristics (pH≤4.08; organic C concentration in the upper mineral layer (g/kg) ≤74.2; total N concentration in the upper mineral layer (g/kg) ≥4.1 and ≤11.13). The three histograms on the right show the mean defoliation in years 2001–2003; the light grey shows the distribution for all plots; the dark grey highlights distribution for the plots selected using the dynamic query tool. The tool provides an immediate feedback when query constraints are modified. The combination of these techniques supports interactive detection of subsets of objects with particular characteristics.

Example 9. Analysis of dynamics of forest fires.

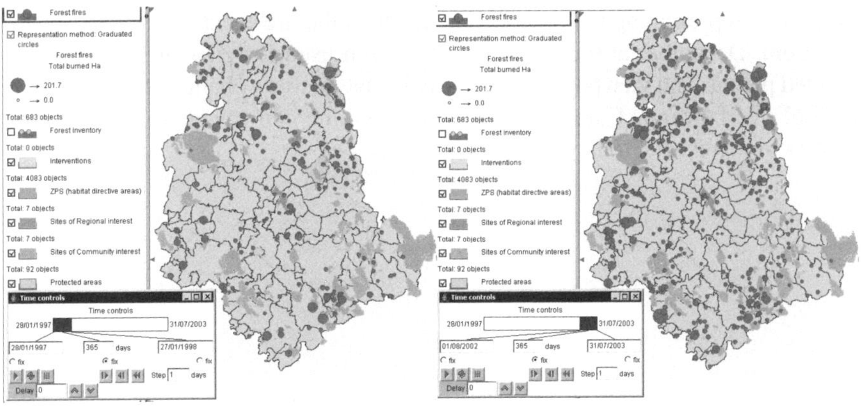

Forest fires in the region of Umbria in Italy are described by their dates and intensity (Example 9). A user-controlled animation of maps demonstrates the dynamics of fire events over the period from 28th January 1997 to 31st July 2003. The intensities of the fire events are represented by circle sizes. These techniques can help to reveal periods of frequent or rare events and periods of numerous fires in specific areas.

Example 10. Detecting characteristic features of spatio-temporal behavior.

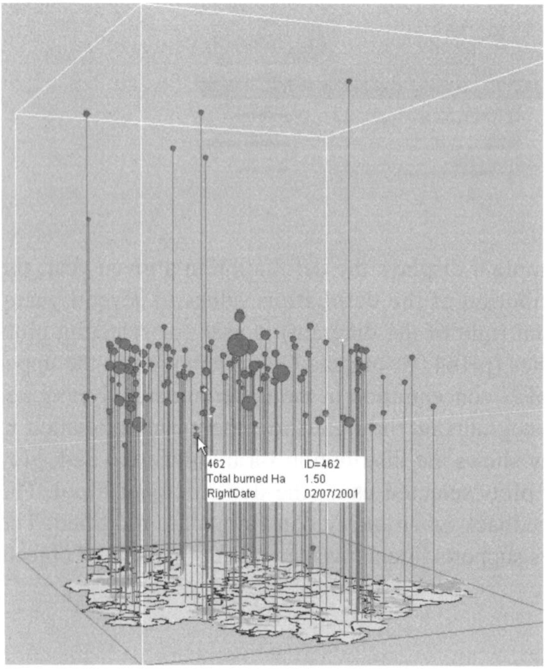

In Example 10, the time of fire events in Umbria has been projected into a third dimension (representing the time of year at which the fire event occurred – January at bottom, December at top). Focusing has been used for viewing events from a selected period. The viewpoint position can be interactively changed for minimizing symbol overlapping. Using these techniques, one can reveal spatio-temporal clusters of fire events, if they exist.

Example 11. Analysis of potentially periodic data.

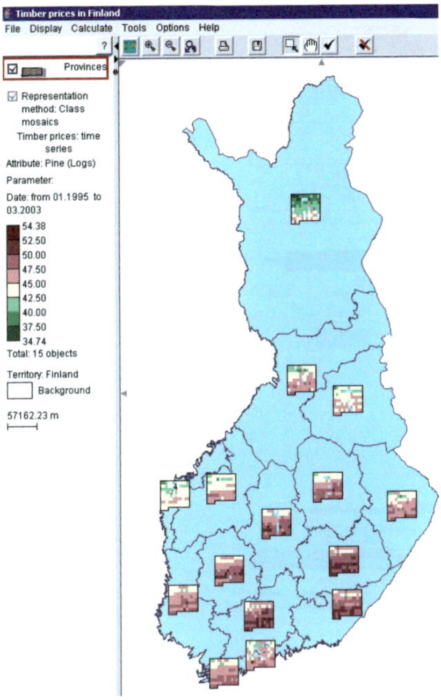

The diagrams shown in Example 11 are designed to represent the complex dynamics of forest prices in Finland's administrative regions (in this case pine log prices in €/m³). The rows of the diagrams correspond to years, and the columns represent the months of one particular year. This representation may reveal any seasonal component of the dynamics (colour patterns) while also demonstrating the general tendency of values over several years.

Example 12. Analyzing non-periodic spatial time series (1).

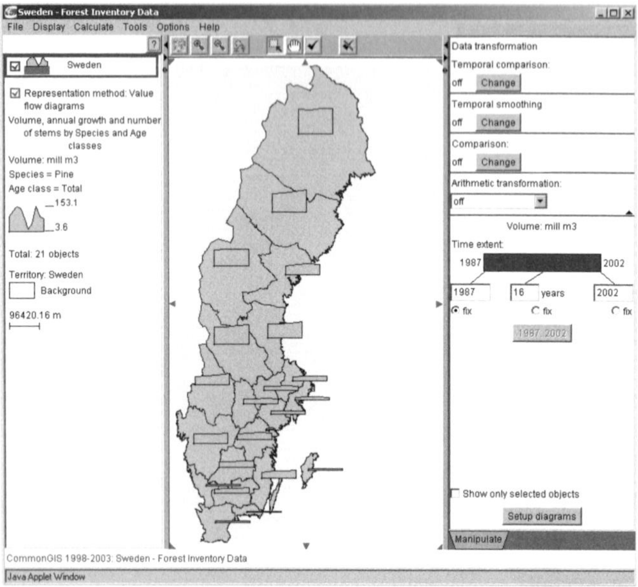

In Example 12, the diagrams represent profiles of annual growth in million m3 (for the tree species 'pine' –all age classes) for administrative regions in Sweden. This map with its diagrams shows the individual dynamics for each region and supports visual grouping of similar dynamics in neighboring administrative regions.

Example 13. Analyzing non-periodic spatial time series (2).

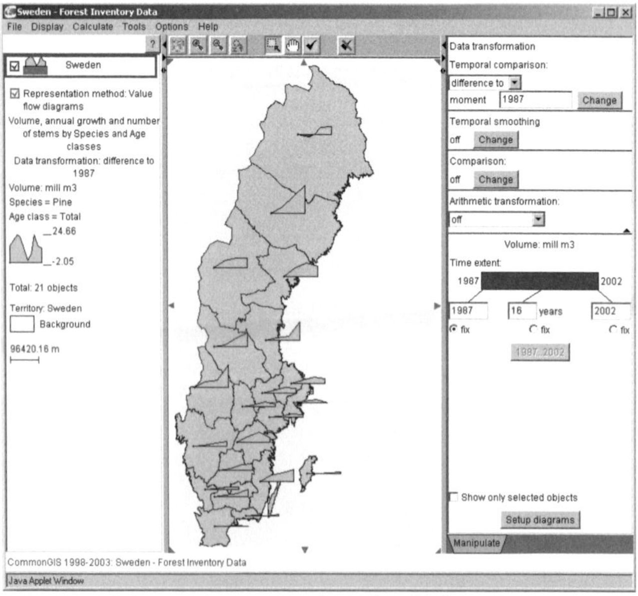

In Example 13, the annual growth in Sweden has been transformed into the increments of the growth since 1987 differentiated by individual regions. The map display emphasizes changes in comparison to a selected time, and can help identification of patterns between regions.

Example 14. Interactive aggregation of multiple raster maps.

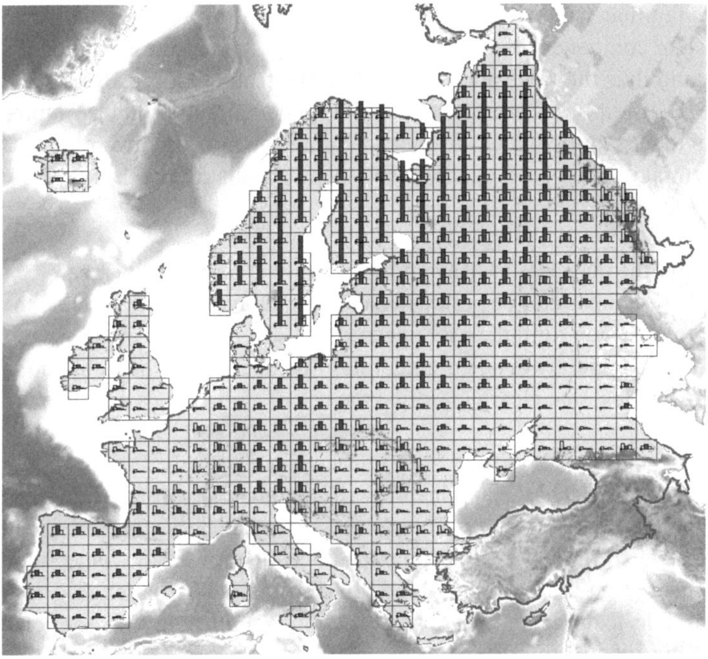

For combining data from multiple raster maps (possibly with different extent and resolution) dynamic aggregation has been used in Example 14. The display of the results (bar charts in this example) automatically changes in response to changing the resolution of the aggregation. This technique is useful for getting an overview and building a composite display of multiple raster maps. The interactive and dynamic properties of the VTK are particularly useful for the analysis of the sensitivity of aggregation. The map shows the proportion of broadleaved, coniferous and mixed forests (left bar, middle bar, and right bar, respectively) within each grid cell.

4.3.4 Evaluation of the visualization toolkit

The evaluation of the VTK was based on the feedback and results of a standardized questionnaire sent to the NEFIS project partners. The questionnaire was developed by the project partner University of Hamburg (UHH), Institute of World Forestry. The data providers of NEFIS were asked to reflect on their experiences when applying their own or other data to the VTK.

Although the results of the questionnaires – due to the small sample size – can only be evaluated as tendencies rather than definitive and objective results, they indicate aspects which should be considered in the improvement of the VTK.

The VTK offers a full range of features which enables analysis and visualization of linked data. To efficiently use and apply relevant features for a particular dataset, users need to familiarize themselves with the VTK. For many of the questionnaire respondents, this was the first encounter with the VTK. As with any software, users will only become familiar with its functionalities by using it. A pre-requisite is that such an EDA system component can be easily understood. It is also essential that a user has a need (and thus the motivation) to apply EDA features for analyzing and visualizing data. That implies that a user is able to generate and produce outputs that fit their individual requirements. In understanding this process, two elements were seen as essential by the questionnaire respondents for making such a software widely used and accepted:

- clearly identify the purposes of use and by whom (e.g. basic or advanced user, scientist, administrator or policy maker, etc.). Only then is it possible to develop features which fulfil the user needs.
- make available sufficiently detailed guidelines/instructions and help functions (including examples) that clearly describe the use of applications and their features.

The question of how users rated the applicability of the functions for analyzing and visualizing a particular dataset, was answered by most respondents as rather difficult (Figure 30). It was stated that all applied functions once identified and understood are attractive to use. Most respondents agreed that it is quite a challenging task to explore all possible features and to identify the most applicable VTK options. It was noted that the VTK outputs need to be clear and have transparent meaning – for example, titles and units must be clearly visible and allocated to the respective dataset and output feature.

Further the VTK offers many features which may rather be of interest for advanced/frequent users, but are of too high complexity for new or occasional users. Close attention will need to be given on clearly delineating potential fields of application for the VTK. Incorporating too many analysis options into one EDA and visualization system may lead to either overwhelming of inexperienced users, and conversely simplifying the system may not meet the high demands of experienced users. A possibile approach could be to: (a) develop a VTK with limited features for those who need a basic set of options only or who are new users; and (b) develop a VTK which includes more advanced features for experienced users.

Another requirement for the VTK use was that users with little or no background in statistics or GIS should be able to use the system easily and retrieve information as demanded. In order to facilitate this, the VTK incorporates automated map design. Automated map design means that:

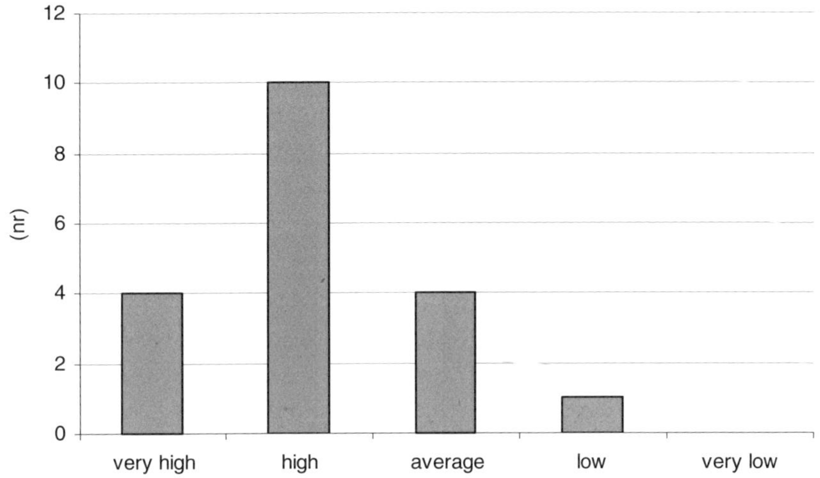

Figure 30. Level of difficulty in understanding the VTK features and applications.

- maps are produced automatically once the command for visualizing a certain dataset is given;
- outputs are available both as data tables and maps where a user can select from different presentation techniques;
- the maps are interactive allowing directly visible manipulations (e.g. zooming, colour scheme adjustment, bar and pie chart editing) but also further exploration with other tools building on the produced base map;
- it allows the selection of statistical displays (box plots, dot plots, scatter plots, etc.) which are then dynamically linked to maps.

One concern was raised by data providers that even if a system does not require highly trained skills in its effective use, there remains the question of fully grasping the processed data, their limitations of applicability for certain analysis and the actual visual result. This is where a step-by-step data exploration is crucial (see section 4.3.1.1). Nonetheless most NEFIS partners considered that the automated map design is appropriate and helpful for users without expert knowledge of GIS for sufficiently mapping and visualizing explicit information (Figure 31).

Despite the current obstacles and future challenges towards implementing the VTK as a key component within a European forest information system, the questionnaire respondents concluded that it provides a solid basis for performing a broad variety of exploration, analysis and visualization tasks as required by various users for their datasets. In order to meet further demands towards effective application of the VTK, it will be essential to analyze in detail the potential user community, their individual requirements, and determine from this analysis what the key components of the VTK are, and how they can be developed.

Reporting will be a main function of any future forest information system at the European scale. Some of the available features of the VTK (e.g. times series,

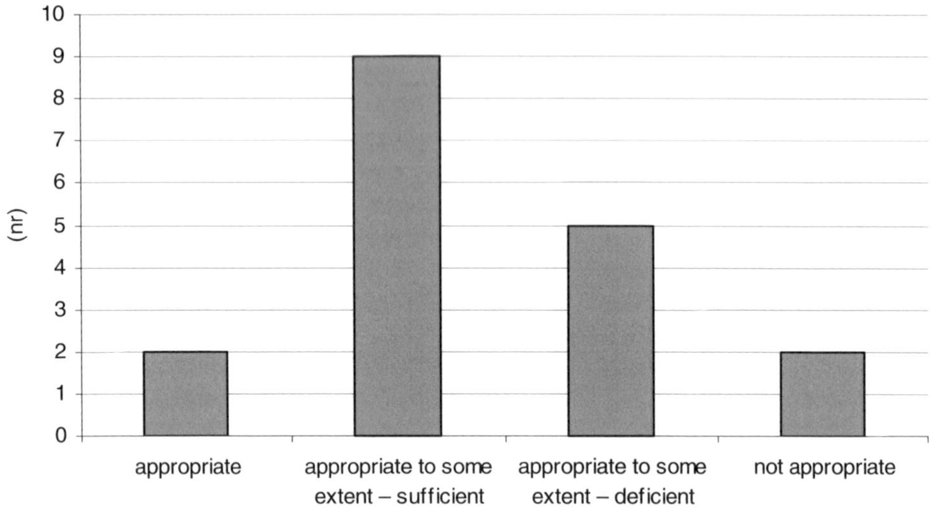

Figure 31. Appropriateness of the automated map design for someone without expert knowledge in GIS.

various forms of mapping and multi criteria decision making) are very useful both for EDA and periodic reporting. However, there will be the need for tailor-made query schemes and output features based on particular reporting obligations (such as required e.g. by the UNECE/FAO, the MCPFE or Eurostat). Such tailored query and reporting schemes (or Web services) may utilise a tool such as the VTK which can provide interactive visualization and EDA functionality. Targeted Web services serving directly organizations, administrations or political/reporting processes can be regarded as a contribution towards putting a European forest information system into practice. It will require close cooperation with those who should be addressed by the services; such investigation was beyond the scope of the NEFIS project.

4.4 Remote search demonstrator

The remote search demonstrator (RSD) provides a mechanism to extract data from a standardized set of distributed data sources, collate the data, present it in a table and convert the data into a form compatible with the EFIS. At least two actors are involved in this process: (1) data users who want to discover, extract and use data; and (2) data providers who have data to share.

The basic mechanism of the RSD is based on User Land Inc's open XML-RPC specification which is described at www.xmlrpc.com/. XML-RPC – *"allows software running on disparate operating systems, running in different environments to make procedure calls over the Internet."* RPC is the acronym for Remote Procedure Call.

Figure 32 shows a schematic diagram of how the XML-RPC functions. A collection of data providers implement XML-RPC servers which listen for XML-RPC client requests. These servers provide standard forest data services. Requests are received from clients for the servers to perform standard data services. These services normally involve data extraction from a database, encoding of the data in XML, transportation of the data across the Internet using HTTP, reception of the XML message by the client, parsing of the XML by the client, reconstruction of the parsed XML into binary data objects, collation of the data objects from the collection of servers into a single result set, followed by preparation of the data into a form required by a specific software application for presentation or analysis.

The Apache Software Foundation's XML-RPC implementation has been used to implement the RSD in Java (Apachc Software Foundation, 2005). The RSD software can be used to run any number of services. Users can locate the remote server services through the resource discovery component of the advanced NEFIS demonstrator by searching for 'RPC' (Figure 33).

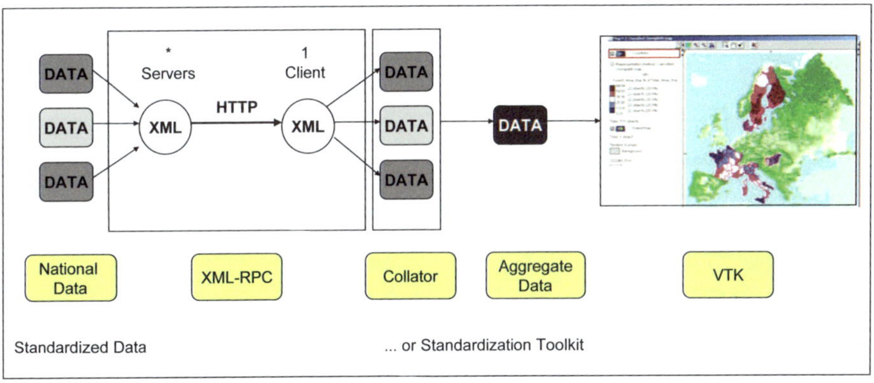

Figure 32. Flow of data from remote server locations using XML-RPC functions (reading from left to right) (after JY Stervinou).

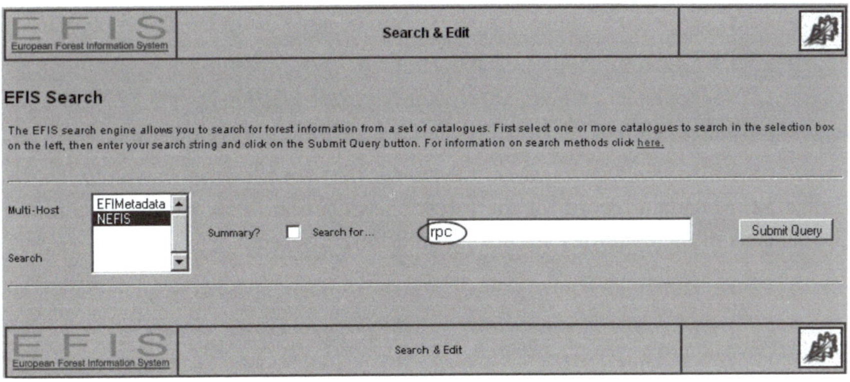

Figure 33. NEFIS example search using the Remote Procedure Call.

EFIS	
Title	Forest land classification in 6 European countries in hectares (created for the EC 5th FP project: Network for a European Forest Information Service, NEFIS)
Creator	European Forest Institute, Finnish Forest Research Institute (Metla), Sveriges Lantbruksuniversitet (SLU), Danish Forest and Landscape Research Institute (FSL), Inventaire Forestier National (IFN), Italian Academy of Forest Sciences (AISF), State Forest Service Hungary • *StreetAddress : Torikatu 34* • *City : Joensuu* • *PostCode : 80100* • *Country : Finland* • *Telephone : +358 13 252 020* • *Fax : +358 13 124 393* • *Email : efisec@efi.fi* • *WebPage : www.efi.fi*
Subject	• silviculture • vegetation types • forest inventory • forest area • land-use classification • other wooded land • land area
	This subset of national forest inventory data (which was created as an example specifically for the NEFIS project) consists of a dataset for six European countries and in particular for forest area and forest land related data at the sub-national level. Also the total land area is included. It should be noted that the data from the six countries namely, Denmark, Finland, France, Hungary, Italy and Sweden, are based on their national definitions and categories of forest and other land areas. Terms used in official translation to English include e.g. wooded areas, forest land, forest area and forestry land. The definitions for the different categories are provided for all countries under the quality report section. The reference dates of the data vary: Denmark, 2000; Finland, 1996-2004; France, 1990 to 2003; Hungary, 2004; Italy, 2000 (based on 1983-1987 inventory results); Sweden, 1998-2002 The data on forest area are retrieved from remote servers i.e. that data is hosted at data providers side. The data are then collated to a data table. The data table (Forest land classification in 6 European countries in hectares) can then be visualised using the NEFIS Visualisation Toolkit. • QualityStatement : Definitions related to forest area. Title of dataset: Forest land classification in 6 European countries in hectares (created for the EC 5th FP project: Network for a European Forest Information Service, NEFIS) --------------- Denmark. Forest is to be or should be planted with tree species which can be developed on the location into high forests with stems, that would stay at least, to a height of 6 meters; the area should be more than 0.5 ha; the area should be wider than 20 meters (at least on average). Source: Study on European Forestry Information and Communication System - Reports on forestry inventory and survey systems Volume 1 and 2, European Commission, 1328 pp. --------------- Finland. The area of forestry land is sub-divided into forest land, scrub land, waste land and other forestry land according to the site productivity. The definitions of the forestry land categories are as follows: (1) Forest land: the potential annual increment of the growing stock is at least 1.0 m3/ha. (2) Scrub land: the potential annual increment of the growing stock is less than 1.0 m3/ha, but at least 0.1 m3/ha. (3) Waste land: unless naturally treeless, the annual increment is less than 0.1 m3/ha. (4)
Publisher	European Forest Institute, Finnish Forest Research Institute (Metla), Sveriges Lantbruksuniversitet (SLU), Danish Forest and Landscape Research Institute (FSL), Inventaire Forestier National (IFN), Italian Academy of Forest Sciences (AISF), State Forest Service Hungary
Contributor	
Date	01/05/2005
Type	EFIS-RPC Server Collection
Format	N/A
Identifier	http://efis.efi.fi:8080/EFISmySQL/XML/EFIS-XML-RPC-Servers.xml
Source	forest.forestAreaByUnits
Language	English
Relation	isVersionOf
Coverage	• Denmark • Finland • France • Hungary • Sweden
Rights	[Access] Public
Audience	educators, government agencies, media, public, researchers, students
Spatial	

Remote Search

Show as XML

Figure 34. Metadata record for the forest classification in six European countries. The datasets of the countries are hosted on different servers.

EFIS European Forest Information System

EFIS-Remote Search Demonstrator Client

EFIS-RSD Client.

Searching 4 servers...

ISO_ID	Country	Area_Name	Total_Area_(ha)	Forest_Area_(ha)	Start_Date	End_Date	Area_Units	NUTS_ID
246000	Finland	Ahvenanmaa	152700.0	61700.0	1996	2004	Ha	FI0
246007	Finland	Etelä-Pohjanmaa	1945400.0	1274000.0	1996	2004	Ha	FI0
246006	Finland	Etelä-Savo	1413700.0	1199400.0	1996	2004	Ha	FI0
246003	Finland	Häme-Uusimaa	1430500.0	942000.0	1996	2004	Ha	FI0
246004	Finland	Kaakkois-Suomi	1072400.0	779900.0	1996	2004	Ha	FI0
246011	Finland	Kainuu	2156700.0	1753700.0	1996	2004	Ha	FI0
246008	Finland	Keski-Suomi	1658300.0	1360700.0	1996	2004	Ha	FI0
246013	Finland	Lappi	9300400.0	5006400.0	1996	2004	Ha	FI0
246002	Finland	Lounais-Suomi	1736500.0	1002400.0	1996	2004	Ha	FI0
246005	Finland	Pirkanmaa	1227100.0	096200.0	1996	2004	Ha	FI0
246010	Finland	Pohjois-Karjala	1778200.0	1490900.0	1996	2004	Ha	FI0
246012	Finland	Pohjois-Pohjanmaa	3529000.0	2410600.0	1996	2004	Ha	FI0
246009	Finland	Pohjois-Savo	1680800.0	1336100.0	1996	2004	Ha	FI0
246001	Finland	Rannikko	1365800.0	823400.0	1996	2004	Ha	FI0
250042	France	Alsace	833150.0	316450.0	1990	2003	Ha	FR42
250072	France	Aquitaine	4180580.0	1797020.0	1990	2003	Ha	FR61
250083	France	Auvergne	2617540.0	706020.0	1990	2003	Ha	FR72
250025	France	Basse-Normandie	1776220.0	167860.0	1990	2003	Ha	FR25
250026	France	Bourgogne	3175020.0	971460.0	1990	2003	Ha	FR26
250053	France	Bretagne	2748380.0	326850.0	1990	2003	Ha	FR52
250024	France	Centre	3953010.0	886080.0	1990	2003	Ha	FR24
250021	France	Champagne-Ardenne	2572160.0	665330.0	1990	2003	Ha	FR21
250094	France	Corse	872600.0	252770.0	1990	2003	Ha	FR83
250043	France	Franche-Comte	1630550.0	691260.0	1990	2003	Ha	FR43

Visualise these data

Figure 35. Extract from the dynamically generated html table, showing the forest area for different administrative areas in Finland and France.

The search locates metadata records that have 'RPC Server Collection' as the data type. Currently, there is one metadata record 'Forest land classification in 6 European countries in hectares' (namely Denmark, Finland, France, Hungary, Italy and Sweden). The search returns total land area and forest area data (in ha) at the sub-national level for these countries (Figure 34). It should be noted that national definitions apply for both forest and other land. The reference dates for the data also may vary. For this demonstration of RPC no attempt has been made to harmonize definitions and thus making data directly comparable (see also section 3.4.2.4).

The datasets for Finland and Sweden were hosted by NEFIS partners on their servers. The datasets for Denmark, France, Hungary and Italy are currently hosted on two separate servers located at the European Forest Institute. By clicking the 'remote search' button, the data are extracted from the remote servers and dynamically compiled into an html table (Figure 35).

These data have been actively linked to the VTK ('visualise these data' button) thus allowing their analysis and visual display (Figure 36).

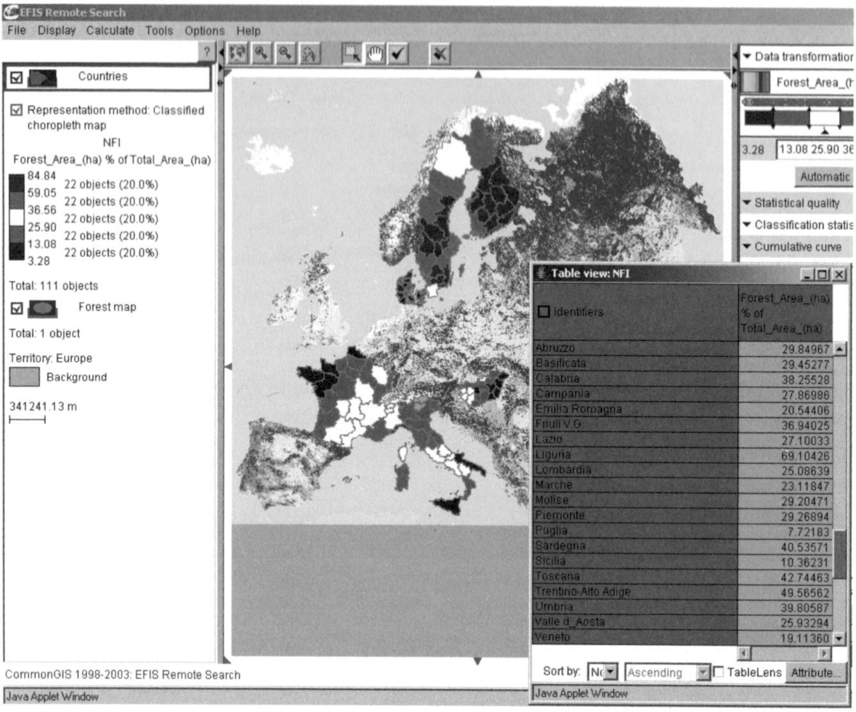

Figure 36. Example of visualizing the forest area data. Display of the percentage of forest area as of the total land area in map format for six European countries . Data for the Italian regions are shown in the table. The background image is the forest area map of Europe (Schuck et al., 2003).

5. A European Forest Information System – The Way Forward

A stable set of Internet standards has enabled the development and growth of the World Wide Web. This has led to the expanding publication of, and easier access to, a vast store of information. Policy- and other decision-makers must attempt to identify and apply the best available knowledge, based on the best available information and data. However, the information is frequently fragmented, information from one source may be incompatible with information from other sources, it may be derived from uncertain sources, and might be of variable or unknown quality. This makes identification and integration of information from different sources at best difficult. In the worst case information may be used for purposes for which it is not suited, and may influence policy- and decision-making.

As outlined in Chapter 1, the improvement and access to environmental information was formally recognized as a priority in 1992 by the United Nations Conference on Environment and Development, Agenda 21, Chapter 40 (UNCED, 1992). Improving access to information and increasing re-use of available information is also a priority for the European Union (EC, 2003).

There are clear benefits of data sharing for the scientific community and society as a whole, and these are comprehensively summarized by (Ball et al., 2004).

> *First and foremost, fully and freely available data promote and reinforce open scientific inquiry, allowing a researcher's conclusions to be validated or refuted by his or her peers. Second, it enables new analyses to be performed, which may lead to novel conclusions. This is especially important in light of the fact that rarely do researchers exploit the full potential of high-throughput datasets upon initial publication. Third, an accumulated body of public data can serve as the basis for new research and new methods of data analysis, and it provides large training and test sets for quality assessment. Fourth, access to public data can provide an excellent teaching resource. Fifth, the accumulation of public data provides all researchers with access to a data set that is larger than one that could ever be constructed by a single laboratory. It is clear that new knowledge and insight can be obtained from analyzing combined datasets, which would never be discovered examining the constituent parts. Lastly, sharing data can prevent unnecessary duplication of effort (though obviously some duplication provides rigor), and the public will benefit from a more rapid pace of scientific discovery that will be the result of decreased duplication and the creative reuse of published data.*

Apart from the altruistic arguments to pursue standardized collection of data and metadata – i.e. that this would allow greater use to be made of our data resources – the idea must be sold to the organizations and individuals that collect and own the data. There are good organizational reasons for the standardized collection of these

Table 13. Organizational benefits of metadata (Wayne, 2005)

Data Archive: Data are the most expensive components of information systems. Metadata is a means of preserving the value of data investments. This is of particular significance to institutions that experience rapid staff changes.
Data Assessment: From a consumer perspective, metadata provides a means to assess available data products. From a producer's perspective, metadata is a means of declaring data limitations and serves as a form of liability insurance.
Data Management: Metadata enables organizations to retrieve in-house data resources by specific criteria for global edits and annual updates.
Data Discovery: Metadata is the primary means of locating available geospatial data resources via the Internet. Metadata is a primary public information resource as it is a non-technical means of presenting technical information.
Data Transfer: Metadata is increasingly used by software systems as a means of properly ingesting data and by analysts as a means of properly displaying data.
Data Distribution: By building metadata in compliance with national and international standards, you can participate in national and international initiatives. Participation promotes your organization and frees staff from answering data inquiries.

data and metadata. The implementation of standards can save money. Initiatives that implemented for example geospatial standards turned out to be more cost efficient as compared to such that relied on in-house proprietary standards (Booz Allen Hamilton, 2005). Organizational benefits of compiling metadata are shown in Table 13.

5.1 Architecture

European forest information systems include vast and ever expanding volumes of information that may be employed by a huge range of potential users and evolving applications. They reside on distributed systems within the context of a rapidly changing socio-technical environment and consist of heterogeneous elements working to differing standards, protocols and natural languages. It is clear that any single monolithic system or 'big-bang' development is bound to fail – if not technically then operationally. It is therefore recommended that any proposed system be an evolving collaboration of communicating systems. The development should take place iteratively employing a component oriented implementation exploiting the current extensive portable web and open-systems technologies. It needs to be based on a well partitioned, scalable architecture within an extensible, service-oriented framework. Development should focus on provision of metadata management facilities and not just for resource discovery. It is argued that normalization, and effective management of metadata and incorporated namespaces will be essential to ensure semantic interoperability. Linked to this is the provision of content management which is central to asset management, dynamic control and delivery. Facilities for maintenance of a tool and standard component repository should be considered. Building upon these elements task and role support, including process and workflow templating and choreography, will facilitate and enhance usability and future perceived success. A central component of the working system will be the provision of high quality publish and subscribe tools.

Four core subsystems were identified:
1. resource and content management;
2. tool and component repository;
3. task and role support; and
4. metadata management.

To ensure the exploitation of highly cohesive, lowly coupled components within what is essentially a distributed heterogeneous system requires a common interchange language and middleware with agreed terminology. Therefore, metadata is not just needed for resource discovery or informed decision support. It also acts as the glue that ensures the semantic interoperability (agent dialogue) of the system. Thus it may be argued that metadata management facilities are the most important of the NEFIS subsystems.

5.2 The adoption of standards

The *Plan of Action of the World Summit on the Information Society* (UN/ITU, 2003) asserts:

> *Standardization is one of the essential building blocks of the Information Society. There should be particular emphasis on the development and adoption of international standards... International standards aim to create an environment where consumers can access services worldwide regardless of underlying technology.*

Decisions must be made concerning the standards needed to support retrieval and interoperability. Those decisions must be made in the light of consensus in other subject areas, and possible EU requirements.

Choosing the standards requires further work, and should involve all interested parties, including users. This would best be achieved by a specialized working group, preferably including specialists who are familiar with interoperability issues, and representatives from other communities within the EU who are aware of emerging regulatory frameworks. This 'community of interest' (COI) could then determine future working structures. As an indication of the work that needs to be done, the following is a non-exclusive list of elements required with examples of the standards or technologies that might be adopted. It is not intended to be a set of recommendations, as there are many issues to be considered before recommendations can be made:

- Top-level metadata format: Dublin Core [already implemented]
- Data syntax: XML [effectively agreed]
- Registration of semantics of shared data elements (word lists etc): ISO 11179 (Information Technology – Metadata Registries)
- Interaction at systems interfaces: WSDL (Web Services Description Language), ebXML (Electronic Business using eXtensible Markup Language), Unified Modeling Language (UML) [partly in use]
- Information discovery: ISO 23950 (Information and documentation – Information retrieval), with profiles SRU/SRW (Search and Retrieve URL/ Web Services)
- Documentation and representation: ISO 19115 (Geographic information – Metadata), ISO 19139 (Geographic information – Metadata – XML schema implementation).
- Place codes: ISO 3166 (Country names and code elements) [in use]
- Extended metadata DTD (Document Type Definition) and XMLS (XML Schema): EAD (Encoded Archival Description), using ISO 8879, Standard Generalized Markup Language (SGML).
- Metadata interoperability: OAI/PMH (Open Archives Initiative Protocol for Metadata Harvesting)
- Semantic structures: RDFS (Resource Description Framework Schema), OWL (Web Ontology Language)
- Envelope (to package all required elements within a document or catalogue entry): METS (Metadata Encoding Transmission Standard)
- Service delivery mechanisms: Web, intranet, offline distribution – these may each require different structuring

One may ask the question "Is metadata needed at all to achieve data interoperability?" There is at present considerable development work in progress in machine recognition and extraction of meaning from text and other data, for example, the NIST ACE (National Institute for Standards and Technology – Automatic Content Extraction) program (NIST, 2004); and the development of ontologies, automatic translation, metadata schemas for ontology and the use of semantics to enhance access to domain knowledge (see for example the FAO Agricultural Ontology Service project –AOS). Awareness should be cultivated of these activities, and it should be ensured that benefits are delivered to the forest sector as production services are developed. At present, however, automated processes still require considerable editorial input to produce workable results. It will be essential for the forest sector community to participate in such developments – e.g. through the COI.

The case of an operational forest information system will require standardized components that will allow the capture and storage of dataset information necessary to make interoperability feasible. These systems will evolve, but, if established standards are adopted, data migration should be relatively straightforward. Initially, the dataset information will be that which has already been gathered or can readily be provided. In the medium term, however, much more metadata will become available as we learn how to extract more meaning from datasets, through a variety of automated and mediated processes. This in turn will allow much deeper searching of the 'semantic web' and make a forest information system a much more useful tool to a much wider community. While working towards practical solutions for a production service based on achievable technology, we need to be aware of and involved in semantic research to ensure that new opportunities can be seized as they arise.

It must be ensured that the systems are understood, publicized and used by the target audiences, and by the current and potential data providers. This largely educational work is best undertaken by the huge 'installed base' of existing information intermediaries, involved in the identification, collection, storage and retrieval of information, in the governmental, commercial, educational and public arenas. The target audience should also be represented in the COI; without their support, future projects and systems are likely to remain hidden. Much of the technical discussion needed in devising interoperable solutions is opaque and unnecessary to the end user and to some extent the data providers; emphasis must be placed on describing the aims and methodology of the projects and systems in language appropriate to the target audience.

The following list summarises the main observations from the NEFIS project relating to standards:

- A forest information system should employ metadata to enhance interoperability in a standards-led process that has long-term sustainability.
- Descriptive metadata is not by itself sufficient for cross-searching.
- Open standards are essential for maximum accessibility.
- Solutions used by other EU agencies should be investigated and where suitable adopted, utilizing systems with a good track record of reliability and can readily incorporate changes to newer technologies as they mature.
- The user should be offered maximum flexibility in retrieval, using mechanisms familiar to the user, but achieving a level of precision appropriate to the data sought.
- Closed, vertical, unscaleable and frequently proprietary information systems that mimic their paper-based predecessors and cannot share information across internal structures are not desireable.
- Services should be developed from a 'customer-centric' viewpoint.
- 'Brand confidence' towards a system needs to be built to ensure that data access on offer corresponds with user needs.
- Clear definitions are needed of the particular types of query the system is aiming to solve, and of the scope of the range of datasets to be included.
- Emerging requirements of the European Interoperability Framework (EIF) should be observed, covering technical, semantic and organizational interoperability (EC, 2004b).
- Approved tools and guidelines need to be developed that address the interoperability of the data and are designed to have a minimum impact on the data providers' business practices and system management.
- Collaborative arrangements with organizations/initiatives working in similar fields should be sought.

- A specialized working group or 'community of interest' (COI) (including users, specialists who are familiar with interoperability issues, and representatives from other communities within the EU who are aware of emerging regulatory frameworks) should be established to advise on important issues and required decisions.
- Existing information intermediaries should be leveraged to ensure the information system is understood, publicized and used by its target audiences and data providers.

5.3 Vocabulary recommendations

Current activities in the terminology work field point towards the development of ontologies for the semantic web. Sound development of ontologies needs to be based on terminology, thesauri and classifications including definitions, relationships and structure. Transforming existing thesauri into ontologies can yield increased precision of semantics particularly for information retrieval purposes. As set out in section 3.3 there are ongoing developments of internationally recognized thesauri which illustrate this trend.

Based on the experiences gained during the activities of vocabulary development within the NEFIS project, the following observations and recommendations are made for advancing vocabulary development in the forestry domain:

- Co-ordinate the various existing vocabulary activities within a Multilingual Forestry Ontology Project, with strategic links to such ongoing ontology framework projects as the FAO Agricultural Ontology Service (www.fao.org/aims/aos.jsp), allowing expression of extensive relationships (beyond the traditional Thesaurus broader/narrower/related terms).
- Establish an Editorial Advisory Group comprising subject experts and information specialists to collate and maintain the forestry ontology.
- Establish practical links with international initiatives concerned with terminology harmonization in forestry and related fields to feed into the ontology maintenance process.
- Produce appropriate authority lists for both field labels and variables within the metadata schema, and standard data definitions that allow deviations from those standards to be clearly identified.
- Tracking the provenance of keywords is a key issue for the development of a controlled vocabulary.
- Encourage use of the Global Forest Decimal Classification as part of the collection of metadata related to forestry resources. The GFDC replaces the former Forest Decimal Classification (FDC) and Oxford System for Decimal Classification for Forestry (ODC). An extensively updated bilingual full version of the English/German edition of the Global Forest Decimal Classification (GFDC) was published in 2006 (Holder et al., 2006). The GFDC is the official expansion of the forestry numbers of the Universal Decimal Classification (UDC).

5.4 Data rights and data rights management

Misuse or abuse of data were a major concern of the data providers. These are issues of fundamental importance (already identified in the EFICS study). A European forest information system should allow role based access rights to the data and metadata to be assigned to particular users.

- The potentials and limitations of data rights management in an implemented and full working information system will be very much linked to the individual needs and requirements set out by the data providers. Where data are stored centrally, access rights should be defined by the data provider.
- Administrative options/restraints towards user communities are fully in the hands of the data providers. Access control and user authentication are challenging technical issues. The solutions will require close cooperation with the data providers in order to fully address the needs of the data providers.
- The responsibility of data access management will need to be fixed through documentation (in the metadata), in particular if datasets are not hosted by the actual data providers.

It is not possible to state categorically whether central server data storage or a distributed network of data sources is the better option. Both approaches have their advantages and disadvantages. The option used by data providers or a network of data providers will depend largely upon individual access policies, and possibly upon technical abilities.

5.5 Geovisualization

On the basis of the experience gained from the NEFIS project, a 'portrait' of an effective forest geovisualization system, encapsulated within the tool repository component, may be outlined.

- It has a well-designed, clear, and user-friendly interface.
- It includes a range of exploratory instruments with complementary capabilities: visual displays supplied with controls for: zooming, focusing, re-arranging, and other display transformations, tools for data transformation such as aggregation or change computing, facilities for querying, filtering, selection, partitioning, and grouping; it also incorporates tools to compute various statistics, proportions, rates, and so on. All these instruments are well integrated and can cooperate in a seamless, organic manner.

- It can recognize which instruments make a minimum combination appropriate to analyse a specific data collection taking into account the nature of the data components (in particular, involvement of space and/or time), relationships between the components, and distributional properties. On this basis, it can configure its user interface so as to hide the underlying complexity. Upon a user's request, it can guide the user through the process of exploration and analysis.
- It does not only allow the user to look at the data from various perspectives and explore various inherent aspects but actively promotes comprehensive analysis by exposing the inherent aspects and prompting the user to consider them.
- It unites exploratory techniques with confirmatory ones so that any finding or hypothesis could be immediately tested.
- It allows the user to record and comment his/her findings immediately after they have been encountered, to come back to earlier findings and recall the history of coming upon them, to organize the records, browse through them and search for particular findings.
- It enables easy and effective reporting of the results of analysis. It supports the selection of relevant information according to the purposes of communication and the role and interests of the intended recipient. It can present the selected information at the desired level of detail and in an appropriate visual form.

5.6 Requirements for international reporting

International reporting requires that national data are comparable to allow holistic assessments about status and trends of relevant concerns. It is well known that nomenclature and classification systems of collected data in forest resource assessments in Europe vary substantially (Päivinen and Köhl, 2005). Data adaptation and conversion to international definitions and requirements are a national priority in order to comply with international commitments such as the Framework Convention on Climate Change (FCCC), the Convention on Biological Diversity (CBD) or the Convention on Combating Desertification (CCD) and comply with reporting obligations towards institutional bodies such as the Food and Agriculture Organization of the United Nations. Thus numerous inititiatives addressing the harmonization of nomenclature and classification systems are ongoing. The aim of the NEFIS project was not to achieve harmonized data comparability, but it rather focused on developing a metadata standard and demonstrating how a functioning system might work.

As forest resource assessments are financed by individual nations, one important aim of countries is to preserve consistent time series based on national systems of nomenclatures which allows backward comparability of national datasests. Delivering national figures according to international standards is often difficult and may even fail because additional assessments or costly data transformations are needed, but are not covered by the available financial and time resources (Köhl et al., 2000). At the moment there are no regulations or agreements on who will cover the costs of harmonization and data conversion. National correspondents are responsible for the transfer and adjustment of national data to international data. Often 'expert guesses' are a common tool to solve this task. Inevitably, these adjustment processes, while increasing the comparability and internal consistency of the international dataset, reduce the accuracy of records by introducing a supplementary source of error (UNECE/FAO, 2000). Various existing international reporting commitments could benefit by linking national datasets in one European service, enabling simpler, faster and more flexible access to data. A fully operable and implemented European forest information system could serve in minimizing the individual national reporting burden, but not in minimizing necessary processes to adjust national data into international data.

Harmonization of data focuses on given systems of nomenclature and data assessments, and seeks to provide definitions that preserve the largest similarities of different national concepts. A straightforward solution to harmonizing national data is the derivation and application of conversion functions. These functions have to take national forest characteristics into account and compensate for the over- or under-estimation of national definitions with respect to the harmonized system of nomenclatures. For some attributes conversion functions may have to be supplemented by additional assessments to guarantee the desired data quality (Köhl et al., 2000). The development of 'conversion toolkits' in the form of software or statistical scripts which can be implemented into national databases to convert national data into international data is seen as a feasible way forward. However, such toolkits have not yet been very far developed. Once available, such toolkits could become integral parts of a European forest information system for international reporting, thus allowing for efficient processing and timely availability of comparable data. In order for a fully operational European forest information system to become a useful tool for harmonized data transfer and reporting, it will be necessary to closely monitor, and where applicable, implement the actions and outputs of various harmonization and streamlining processes as they are addressed by various groups:

- Expert Meeting on Harmonizing Forest related Definitions for Use by Various Stakeholders;
- Collaborative Partnership on Forests (CPF) –Task Force on Streamlining Forest-Related Reporting;
- COST Action E 43 'Harmonisation of National Forest Inventories in Europe: Techniques for Common Reporting';
- the Minsterial Conference on the Protection of Forests in Europe (Expert level working groups);
- the UNECE/FAO Team of Specialists on 'Monitoring forest resources for Sustainable Forest Management in the UNECE Region'.

A forest information system should support and enable more efficient delivery of data by national correspondents generation of data for specific needs by different constituencies such as FAO/UNECE (FAO, 2006) or MCPFE (C&I). It should allow direct feeding of data to, for example the UNECE/FAO databases, and automatic consistency checks for the data supply. In doing this, it could enhance the value of existing data for different user groups without adding to the workload of national correspondents and data storing institutions. The potential for how far the current system can serve as a reporting tool depends not only on the applicability and the usability of the system, but also on the amount, content and format of the provided datasets. That means it depends very much on those who will provide data to such a system. To define the need and purpose of the system for reporting requires an analysis of the central reporting requirements – at the global, regional, national and sub-national levels. It will be essential that an investigation is performed on who will be the consumer of the provided, linked, analyzed and visualized data, and what information they will require. By collection of this background information, the functionalities and applications of the system and the content and format of provided data can be developed further, adapted and supplied in their required form. Reporting requirements are very complex and subject to change. It is therefore essential to identify and classify the range of specific requirements and analyze how far these can be efficiently addressed. The knowledge of networking structures, information flows and information patterns are fundamental to understanding in how far a European forest information system has the potential to place itself within the broad context of information demands and supply. It also determines in how far it can act and serve for reporting, and policy- and decision-making. The outcomes of NEFIS will in particular support the Joint Research Centre of the European Commission (Land Management Unit) in coordinating and supervising the development of a European Forest Information and Communication Platform (EFICP). The EFICP, once fully operational, will represent one component of a data portal for the European Union and thus a broad variety of user communities.

5.7 Final comment

There is a need, and indeed an obligation on the part of Member States of the EU, for development of systems and services that enables description, identification, evaluation and analysis of forest-related information in Europe. In the same way that the development of the Web was enabled by a stable set of Internet standards and protocols, the adoption of stable standards developed by communities to meet the needs of those communities will lead to the development of varied systems and services. Stable and flexible data and metadata standards will enable more effective data identification and ensure a high level of compatibility (interoperability) between data sources. They will thus allow for the development of a variety of as yet unimagined services and systems. Much collaborative work will be needed, but the pathways exist if the political will is present.

References

Andrienko, G., Andrienko, N. and Voss, H. 2003. GIS for everyone: the CommonGIS project and beyond. In: M. Peterson (ed.). Maps and the Internet. Elsevier, Amsterdam. Pp. 131–146.

Andrienko, N. and Andrienko, G. 2006. Exploratory Analysis of Spatial and Temporal Data. Springer-Verlag, Berlin Heidelberg. 703 p.

Apache Software Foundation. 2005. About Apache XML-RPC. http://ws.apache.org/xmlrpc/

Ball, C.A., Sherlock, G. and Brazma, A.. 2004. Funding high-throughput data sharing. 22(9): 1179–1183.

Batschi, W.-D., Felluga, B., Legat, R., Plini, P., Stallbaumer, H. and Zirm, K.L. 2002. "SuperThes": A New Software for Construction, Maintenance and Visualisation of Multilingual Thesauri, EnviroInfo, Wien, Austria. Pp. 125–132.

Booz Allen Hamilton, 2005. Geospatial Interoperability Return on Investment Study, NASA Geosciences Interoperability Office (GIO). 80 p. http://gio.gsfc.nasa.gov/docs/ROI%20Study.pdf

CCFM. 2004. National Forest Information System (CCFM-NFIS) – An Overview, Canadian Council of Forest Ministers. CCFM-NFIS Project Office. 7 p. http://nfis.org/about/index_e.shtml

CEN. 2003a. Guidance material for mapping between Dublin Core and ISO in the Geographic Information Domain. CWA 14586, Comité Européen de Normalisation (CEN). 98 p.

CEN. 2003b. Mapping between Dublin Core and ISO 19115, "Geographic Information – Metadata". CWA 14857, Comité Européen de Normalisation (CEN). 69 p. ftp://cenftp1.cenorm.be/PUBLIC/CWAs/e-Europe/MMI-DC/cwa14857-00-2003-Nov.pdf

CompTIA. 2004. European Interoperability Framework: ICT Industry Recommendations (White Paper), Computing Technology Industry Association. 40 p.

COSTE43. 2004. Harmonisation of National Forest Inventories in Europe: Techniques for Common Reporting. Memorandum of Understanding (entered into force on June 10, 2004). http://www.metla.fi/eu/cost/e43/cost-e43-mou.pdf. 17 p.

CSD. 1997. Fifth Session. 7–25 April 1997. Report of the Ad Hoc Intergovernmental Panel on Forests on its Fourth Session. 20 March 1997 (E/CN.17/1997/12). Commission on Sustainable Development

Davenport, T. and Prusak, L. 2000. Working Knowledge: How Organizations Manage What They Know. Harvard Business School Press, 224 p.

DCMI. 2004. Dublin Core Metadata Element Set, Version 1.1. Dublin Core Metadata Initiative. http://dublincore.org/documents/2004/12/20/dces/

DCMI. 2005a. DCMI Glossary. Dublin Core Metadata Initiative. http://dublincore.org/documents/usageguide/glossary.shtml

DCMI. 2005b. Metadata Terms. Dublin Core Metadata Initiative. http://dublincore.org/documents/2005/01/10/dcmi-terms/

EC. 1989. Regulation (EC) No 1615/89 of 29 May 1989. Establishing a European Forestry Information and Communication System (EFICS). Official Journal, L 165 , 15/06/1989 P. 0012 0013.

EC. 1997. Study on European Forestry Information and Communication System. Report on forestry inventory and survey systems. Vol.1, 2, European Communities. 1328 p.

EC. 1999. Regulation (EC) No 1257/1999 of 17 May 1999 on support for rural development from the European Agricultural Guidance and Guarantee Fund (EAGGF) and amending

and repealing certain Regulations. Official Journal of the European Communities, L 160/80: 23 p.
EC. 2002. eEurope 2005: An information society for all. Action Plan. 21/22 June 2002. 23 p. http://europa.eu.int/information_society/eeurope/2005/index_en.htm
EC. 2003. Directive 2003/98/EC of the European Parliament and of the Council of 17 November 2003 on the re-use of public sector information. Official Journal of the European Union, L345: 90-96.
EC. 2004a. eEurope 2005 Mid-term Review, Commission of the European Communities. 12 p.
EC. 2004b. European Interoperability Framework for Pan-European eGovernment Services. Interoperable Delivery of European eGovernment Services to public Administrations, Businesses and Citizens (IDABC). European Commission.
EC. 2004c. Proposal for a Directive of the European Parliament and of the Council establishing an infrastructure for spatial information in the Community (INSPIRE). COM(2004) 516 final. 2004/0175 (COD), 23.7.2004, Brussels, Belgium. 32 p. http://inspire.jrc.it/sdic_call/EN.pdf
EC. 2007. Directive 2007/2/EC of the European Parliament and of the Council of 14 March 2007 establishing an Infrastructure for Spatial Information in the European Community (INSPIRE). Official Journal, L 108, Vol. 50. 14 p.
EEA. 2003. EEA metadata form for spatial datasets & EEA metadata standard for geographic information (EEA-MSGI v.1.1), Version 1.2, European Environment Agency. 7 p. http://www.eionet.europa.eu/gis/geographicinformationstandards.html
Endres, A. and Rombach, D. 2003. A Handbook of Software and Systems Engineering, Pearson. Addison-Wesley Professional. 352 p.
Eurostat. 2000. Forestry statistics Data 1995-98. Statistical Document. Theme 5. L15252, European Communities, Luxembourg. 148 p.
Eurostat. 2003. Forestry statistics – Data 1990–2002 – Pocketbook, European Communities, Luxembourg. 72 p.
FAO. 2002. Proceedings. Second expert meeting on harmonizing forest-related definitions for use by various stakeholders, Food and Agricultural Organization of the United Nations, Rome, Italy. http://www.fao.org/documents/show_cdr.asp?url_file=/DOCREP/005/Y4171E/Y4171E00.HTM
FAO. 2004. Global forest resources assessment update 2005. Terms and definitions. (Final version). Food and Agriculture Organization of the United Nations, Rome, Italy
FAO. 2005. Proceedings: Third expert meeting on harmonizing forest-related definitions for use by various stakeholders. Food and Agriculture Organization of the United Nations; Rome, Rome, 17–19 January 2005. 148 p.
FAO. 2006. Global Forest Resources Assessment 2005: Progress towards sustainable forest management. FAO Forestry Paper 147. Food and Agricultural Organization of the United Nations, Rome, Italy. xxvii+320 p.
FAOSTAT. 2006. ForesSTAT. http://faostat.fao.org/site/381/default.aspx
FGDC. 1998. Content Standard for Digital Geospatial Metadata (Version 2.0) (FGDC-STD-001-1998). Federal Geographic Data Committee. http://www.fgdc.gov/standards/standards_publications/index_html
Foster, I. and Kesselman, C.. 2005. The Grid 2: Blueprint for a New Computing Infrastructure. The Morgan Kaufmann Series in Computer Architecture and Design. Morgan Kaufmann. 748 p.
GFIS. 2005. Development of a Global Forest Information Service (GFIS) as a CPF Initiative. GFIS Working Paper 1. Global Forest Information Service. 6 p.
Glass, R.L. 2003. Facts and Fallacies of Software Engineering. Addison-Wesley Professional. 224 p.

GlobusAlliance. 2005. Grid Computing Specifications. http://www.globus.org/

Gold, S. 2003. The development of European forest resources, 1950 to 2000: a better information base. Geneva Timber and Forest Discussion Paper(ECE/TIM/DP/31). viii + 98 p.

Holder, B., Saarikko, J. and Voshmgir, D. 2006. Global Forest Decimal Classification (GFDC). IUFRO World Series Volume 19. 338 p.

ICHNET. 2005. Interoperability Clearinghouse Glossary of Terms. Interoperability Clearinghouse. http://www.ichnet.org/glossary.htm

IDA. 2003. Linking up Europe: the Importance of Interoperability for eGovernment Services. Commission Staff Working Paper, Interchange of Data between Administrations (IDA). Commission of the European Communities. 23 p.

IMRC. 2004. Dublin Core Audience sub-group: Government of Canada audience scheme – Final version. Information Management Resource Centre (Canada). http://www.tbs-sct.gc.ca/im-gi/mwg-gtm/aud-aud/docs/2003/schemfinal/schemfinal_e.asp.

INSPIRE. 2002. INSPIRE Architecture and Standards Position Paper. Architecture And Standards Working Group. P.C. Smits, U. Düren, O. Østensen, L. Murre, M. Gould, U. Sandgren, M. Marinelli, K. Murray, E. Pross, A. Wirthmann, F. Salgé and M. Konecny. JRC-Institute for Environment and Sustainability, Ispra, Italy. 64 p. http://inspire.jrc.it/reports/position_papers/inspire_ast_pp_v4_3_en.pdf

Intergraph_Corporation. 2003. CSDGM to ISO 19115 Element Crosswalk. http://www.fgdc.gov/metadata/geospatial-metadata-standards/

ISO. 1991. ISO/IEC 9126. Information technology – Software product evaluation – Quality characteristics and guidelines for their use. International Organization for Standardization, International Electrotechnical Commission

ISO. 2003a. ISO 15836:2003. Information and documentation – The Dublin Core metadata element set, International Organization for Standardization. 14 p. http://www.niso.org/international/SC4/n515.pdf

ISO. 2003b. ISO 19115:2003. Geographic Information – Metadata, International Organization for Standardization. 140 p.

ISO. 2005. ISO/IEC 11179: Information Technology – Metadata Registry (MDR) standard. http://metadata-standards.org/11179/. Reference date 08/11/2005

JRC. 2002a. European Forest Information System (EFIS) – Executive Summary. Project number: 17186-2000-12 F1ED ISP FI. European Commission – Joint Research Centre. 17 p.

JRC. 2002b. European Forest Information System (EFIS) – Final Report. Project number: 17186-2000-12 F1ED ISP FI. European Commission – Joint Research Centre. 146 p.

Kammersgaard, J. 1990. Perspectives on Human-Computer Interaction. In: J. Preece and L. Keller (eds.), Human-Computer Interaction. Prentice-Hall. Pp. 42–64.

Kennedy, P., Folving, S., Munro, A., Päivinen, R., Schuck, A., Richards, T., Köhl, M., Voss, H. and Andrienko, G. 2004. European Forest Information System EFIS – a step towards better access to forest information. In: P. Corona, M. Köhl and M. Marchetti (eds.), Advances in forest inventory for sustainable forest management and biodiversity monitoring. IUFRO Conference 'Collecting and Analyzing Information for Sustainable Forest Management and Biodiversity Monitoring With Special Reference to Mediterranean Ecosystems. Palermo, Sicily (Italy). 4–7 Dec. 2001. Kluwer. Pp. 295–310.

Köhl, M. 2006. Forest information systems. In: Shao-Guofan and K.M. Reynolds (eds.), Computer Applications in Sustainable Forest Management. Managing Forest Ecosystems. Vol. 11.

Köhl, M., Traub, B. and Päivinen, R. 2000. Harmonisation and Standardisation in Multi-National Environmental Statistics – Mission Impossible? Environmental Monitoring and Assessment, 63(2): 361–380.

Kruchten, P.B. 1995. The 4+1 View Model of Architecture. IEEE Software, 12(6): 42–50.
Lund, B., Hammond, T., Flack, M. and Hannay, T. 2005. Social Bookmarking Tools (II): A Case Study – Connotea. D-Lib Magazine, 11(4).
MCPFE. 2003a. Relevant definitions used for the improved pan-European indicators for sustainable forest management, Liaison Unit, Vienna. 21 p.
MCPFE. 2003b. The state of Europe's forests 2003. The MCPFE report on sustainable forest management in Europe, Jointly prepared by the MCPFE Liaison Unit Vienna and UNECE/FAO. 126 p.
Melnik, S. and Decker, S.. 2000. A Layered Approach to Information Modelling and Interoperability on the Web, Proc. ECDL'00 Workshop on the Semantic Web, Lisbon, Portugal.
NIST. 1993. Reference Model for Frameworks of Software Engineering Environments. Government Printing Office. Prepared jointly by National Institute for Standards and Technology and the European Computer Manufacturers Association (ECMA), Washington, DC, USA.
NIST. 2004. ACE – Automatic Content Extraction. http://www.itl.nist.gov/iad/894.01/tests/ace/
OASIS – Organization for the Advancement of Structured Information Standards. http://www.oasis-open.org/home/index.php
O'Dell, C. and Jackson Grayson Jr., C. 1998. If We Only Knew What We Know: the Transfer of Internal Knowledge and Best Practice. The Free Press. 256 p.
OMG. 2005a. CORBA/IIOP specification. Object Management Group. http://www.omg.org/technology/documents/formal/corba_iiop.htm
OMG. 2005b. The UML specification. Object Management Group. http://www.omg.org/technology/documents/formal/uml.htm
Päivinen, R., Iremonger, S., Kapos, V., Landis, E., Petrokofsky, G., Richards, T. and Schuck, A. 1998. Better access to information on Forests, International Consultation on Research and Information Systems in Forestry. Proceedings. An Austrian and Indonesian initiative in support of the programme work of the Intergovernmental Forum on Forests. Federal Ministry of Agriculture and Forestry, Austria, Gmunden, Austria, September 1998. Pp. 113–131.
Päivinen, R. and Köhl, M. (eds.). 2005. European Forest Information and Communication System (EFICS). EFI Technical Report No. 17. European Forest Institute. 199 p.
Pressman, R.S. 2005. Software Engineering: A Practitioner's Approach. McGraw-Hill. 896 p.
Rennolls, K. 2005. An Architectural Design for a Forest Information and Communication Portal, Int. Workshop on Forest and Environmental Information and Decision Support Systems (FEIDSS'05). In: Proceedings of the 16th International Conference Database and Expert Systems Applications. IEEE Computer Society Press, Copenhagen, Denmark. 22–26 August 2005. Pp. 684–688.
Rennolls, K., Lee, F., Ibrahim, M. and Fedorec, A. 2003. "Web Services" and NEFIS: Part 1. NEFIS Technology Review
Schelhaas, M.J. 2003. Database on Forest Disturbances in Europe (DFDE) – Technical Description, EFI Internal Report 14. European Forest Institute. 44 p.
Schelhaas, M.J., Varis, S., Schuck, A. and Nabuurs, G.J. 2006. EFISCEN – The European Forest Information Scenario Model. http://www.efi.fi/projects/efiscen/. European Forest Institute, Joensuu, Finland.
Schuck, A., Andrienko, G., Andrienko, N., Folving, S., Köhl, M., Miina, S., Päivinen, R., Richards, T. and Voss, H. 2005. The European Forest Information System – an Internet based interface between information providers and the user community. Computers and Electronics in Agriculture, 47(3): 185–206.

Schuck, A. and Green, T. 2005. Sharing Forest Information – Users United by Improving Networks. EFI News, 13(3): 5–7.
Schuck, A., Päivinen, R., Häme, T., Van Brusselen, J., Kennedy, P. and Folving, S. 2003. Compilation of a European forest map from Portugal to the Ural mountains based on earth observation data and forest statistics. Forest Policy and Economics. [Diamonds from the European Forest Institute. Evaluation of 10 Years Research in European Forestry Issues.], 5(2): 187–202.
Schuck, A., Van Brusselen, J. and Andrienko, G. 2004. The European Forest Information Demonstrator – EFIS. Querying the Forests of Europe. GeoInformatics 7(2): 42–45.
Tudhope, D. 2004. Semantic Terminology Services: Experiences from the FACET Project, DELOS Workshop, Lund, June 2004.
Tukey, J.W. 1977. Exploratory Data Analysis. Addison-Wesley Series in Behavioral Science, No. 29. Addison-Wesley, Reading MA. 688 p.
UN/FAO. 2001. Forest Resources Assessment 2000. Main Report, Food and Agriculture Organization of the United Nations, Rome, Italy. 479 p.
UN/ITU. 2003. World Summit on the Information Society. Plan of Action. WSIS-03/GENEVA/DOC/5-E. 12 December 2003., United Nations/International Telecommunication Union. 13 p. http://www.itu.int/dms_pub/itu-s/md/03/wsis/doc/S03-WSIS-DOC-0005!!MSW-E.doc
UNCED. 1992. Agenda 21, Chapter 40 Information for Decision Making, United Nations Conference on Environment and Development
UNECE. 2005. Terms of reference and rules of procedure of the United Nations Economic Commission for Europe (UNECE). http://www.unece.org/oes/about/terms.htm
UNECE/FAO. 2000. Forest Resources of Europe, CIS, North America, Australia, Japan and New Zealand: Main Report. 445 p.
Vandenbroucke, D. 2005. Spatial Data Infrastructures in Europe: State of play Spring 2005. Summary report of Activity 5 of a study commissioned by the EC (EUROSTAT & DGENV) in the framework of the INSPIRE initiative, Spatial Applications Division Leuven (SADL), K.U.Leuven. 32 p.
VanderWal, T. 2005. Folksonomy definition. vanderwal.net. http://www.vanderwal.net/random/entrysel.php?blog=1750
Voss, H. 2003. Current status of the standardization in the GI sector. http://www.nefis-kb.info/. 27 p.
W3C, World Wide Web Consortium. http://www.w3.org/
W3C. 2005. Web Services Specifications. http://www.w3.org/2002/ws. Reference date 3-10-05. World Wide Web Consortium
Wardle, P., Van Brusselen, J., Michie, B. and Schuck, A. 2003. Forest products statistical information systems of EU and EFTA. EFI Research Report No. 16. Brill, Leiden-Bosten, Netherlands, 164 p.
Wayne, L. 2005. Institutionalize metadata before it institutionalizes you http://www.fgdc.gov/metadata/metadata-publications/institionalize-metadata
WS-I, Web Services Interoperability Organization. http://www.ws-i.org/